MW00478696

Advance Praise for *ForkFight!*

"The key to Mark's success was his ability to inspire and infect each new restaurant team with a culture that separated those restaurants from both chain and independent restaurants across the country."

—Rick Federico, retired Chairman,
P.F. Chang's China Bistro

"You would think Mark is thirty-five years old with his energy, enthusiasm, wide eyes, hands flailing, and chatter about all the irons he has in the fire. Mark has never slowed down and he transcends generations of restaurateurs and brands which have come and gone."

—Chef Nevielle Panthaky,
Vice President of Culinary, Chipotle

"I can't say exactly for sure but I'd say that Mark's fingerprints are on about 75 percent of the tacos on the Velvet Taco menu. He will undoubtedly be forgotten as the brand continues to grow but there are a select few of us who will forever know just how integral he was to the success of the brand and of all its industry recognition."

—Chef John Franke, former Corporate
Chef, Front Burner Restaurants and current
CEO, Franke Culinary Consulting

"The future of the restaurant business depends on the Mark Brezinskis of the world to tell us where we need to go."

—Bob Sambol, Founder/Owner,
Bob's Steak & Chop House

"Many creative guys are working on renovation, they take something that exists, polish it, and make it look new. Not Mark. He has the capability to create a new blue ocean."

—Christophe Poirier, Chief Brand Officer,
New Business Development, Pizza Hut Global

"Most of us see the world for what it is and do our best to fit in, but there are those select few who start with an idea and never mind that it doesn't already exist. In the world of food and hospitality, Mark is one of those rare visionaries."

—Micky Pant, Former CEO, YUM!
Restaurants International

"As unknown as Anthony Bourdain before publishing *Kitchen Confidential,* Mark Brezinski's *ForkFight!* takes you behind the scenes of visioning, creating, financing, and opening the restaurants you eat in today. It's a world most didn't know existed. It's a messy world. And in many cases it's one that not many can go through financially and psychologically intact. I have lived this industry for decades and learned something new about it in every chapter of the book. Informative, entertaining, and filled with life lessons, *ForkFight!* should yield this humble individual the credit he so rightfully deserves for his contributions to the industry."

—Lane Cardwell, Former CEO, Boston
Market, Former President, P.F. Chang's China
Bistro, Former CEO, Eatzi's Market & Bakery

FORK FIGHT!

Whisks, Risks, and Conflicts Behind the Restaurant Curtain

FORK FIGHT!

Whisks, Risks, and Conflicts
Behind the Restaurant Curtain

Mark H. Brezinski
with Mark Stuertz

Post Hill
PRESS

A POST HILL PRESS BOOK
ISBN: 978-1-63758-750-8
ISBN (eBook): 978-1-63758-751-5

ForkFight!:
Whisks, Risks, and Conflicts Behind the Restaurant Curtain
© 2023 by Mark H. Brezinski
All Rights Reserved

Cover art and design by Amber Brown

Post Hill Press
New York • Nashville
posthillpress.com

Published in the United States of America
1 2 3 4 5 6 7 8 9 10

FORKFIGHT! is dedicated to everyone who has ever believed in me and the restaurants I helped create and operate, including investors, staff, executives, friends, family, consumers, and other believers. It takes intestinal fortitude, a thick skin, and perseverance beyond what anyone can ever tell you need to make this happen.

FORKFIGHT! is also dedicated to my parents: Henry and Betty Brezinski. My dad wasn't always a believer. I honestly don't think he ever learned how. But he supported me in his own way by teaching me a work ethic and the ability to enjoy the fruits of labor that became a part of my own fabric. No one I've known has ever worked harder, and few have displayed the generosity that he was legendary for.

My mother was a tireless believer for as long as she was alive. And she was my source of creativity, sensitivity, and perspective. This book is not about them, but it wouldn't have been possible without them. Neither is ever far from my work, and my appreciation for their lessons is boundless.

Finally, *FORKFIGHT!* is dedicated to everyone who chooses to support the restaurant industry; its workers, vendors, and investors at all levels. We're mostly an industry of people-pleasers, determined to create better ways for you to dine, celebrate, and escape. We may not always show it, but at the heart of our work is a love of discovering ways to persuade you to go out and enjoy a meal with us.

Your choices in the US are the most varied and abundant of any country in the world. Our hospitality industry employs tens of millions and is ever transforming the trends of what and how we eat. Thank you on behalf of all of us who have ever dedicated our lives to making your lives a little bit better through the boundless joys of food and beverage. It's because of you that we do what we do.

Table of Contents

Author's Note

The views and stories shared in FORKFIGHT! *are based on my own memories of the experiences I have had over my career. These are not the views of any of the corporations or people I worked with over that time and any inaccuracies are solely my own. I have done my best to recall and write to the best of my recollection.*

Foreword

I had heard about Mark Brezinski from Chef Floyd Cardoz, my mentor with whom I had worked in New York City. Floyd shared that he was unable to consult with Mark on a project he was working on and wanted to know if I might be interested in pursuing it with him. At the time, I was living in Southern California with my wife Michelle, also a chef, and we were enjoying the beauty of San Diego and all that comes with coastal living.

I was curious, so I checked out Pei Wei Asian Diner, the successful concept Mark had just exited. Combined with what Floyd had shared with me, his new venture seemed to make perfect sense. His vision combed the wonderful cuisines of South Asia with the aspiration of elevating these exotic and vibrant flavors to a national level. It was an enticing proposition.

Mark contacted me from Dallas, and we decided to meet. He came out to the restaurant I oversaw just north of San Diego and brought along his fellow Pei Wei partner Mo Bergevin. I composed a creative tasting menu for them, and we hit it off. He offered me a position on the spot.

We opened Bengal Coast—this taste of "the other Asia" in late 2007. During my time with Mark, spanning some three years, I was inspired by his creativity and fierce dedication to his vision. His passion for educating, inspiring, and connecting with staff, guests, investors, and vendors was awe-inspiring. Sure, we had our disagreements, as creative people often do. But Mark is Mark, stubborn as fuck, loyal to his followers, emotional yet guarded.

His pet peeves included the aromas of fish sauce hitting the wok, not greeting guests at the door within thirty seconds, and experimenting too much with traditional foods and obscure flavors. The cuisine at Bengal Coast was exceptional, and it was a bold idea that was far ahead of its time. The great financial crisis of 2009 thwarted its survival, but Mark did everything he could to keep it alive at great cost to his personal and financial well-being (which I am sure is revealed in these pages).

Mark was always resourceful and forward-looking, figuring out ways to keep Bengal open for one more shift, one more day, one more week. When he finally went bankrupt after all these heroic efforts, I was shocked, stunned, and saddened. He looked me in the eye one night and said, "It's just money, Nevielle. I'll be fine."

I have kept in touch with Mark over the years but have not been able to keep up with his exploits. You would think he was thirty-five years old, what with his energy, enthusiasm, wide eyes, flailing hands, and ceaseless chatter about all the irons he has stoking in the fire. Mark has never slowed down, and his spirit transcends the generations of restaurateurs and concepts that have come and gone.

He is truly an enigma, and he will, in his own right, be remembered for his creativity and passionate love for this crazy business. Servers to the line! Pick up on Jungle Curry, Malai Kebab, and Rendang Steak Salad!

—*Nevielle Panthaky, vice president of culinary at Chipotle Mexican Grill and former executive chef and general manager of Bengal Coast*

Introduction

It's important to get it out there right up front so that there's no confusion: I am no genius, clairvoyant, or saint. I've never pretended to be any of those things. And I've never aspired to be any of them either. My flaws, though perhaps not obvious, are very iceberg-like—unassuming peaks supported by treacherous footings beneath the surface.

Throughout my many years in the restaurant business, I have struggled and failed. I have gambled and won. I have tasted major windfalls and have swung and missed so hard I corkscrewed myself (and my marriages) deep into the mud.

Make no mistake. This is not a "how to" book or a guide to achieving unimagined success. God knows there is a plethora of that kind of pulp out there to choose from. Instead, you're about to read a compilation of life experiences, trials, and tribulations that are intended to inspire, entertain, and amuse. Think of it as an insider's look into the field of restaurant concept creation and management.

The restaurant industry is (or at least it was before COVID-19) the second largest employer in the United States after the government. This look into the all-consuming industry is not always flattering and is often NC-17 rated. It's a field filled with quirky—often batshit crazy—characters who have probably figured into a bite or two of food you've enjoyed over the years. It's an industry that has become a default employer of Hollywood dreamers, wannabe rock stars, the uneducated, and immigrants yearning for opportunity.

It's a field that has the heady pastiche of glamour that almost everyone wants to talk about, invest in, and otherwise become involved with—mostly because of the seductive cosmopolitan social currency it so often expends. It's a show unlike any other. It's relentless and hardly ever closes. It's a show where people spend whole paychecks before descending into their worst depths, a show that few understand, but all want to be a part of.

My first job in the business was flipping burgers on a real charcoal-fired grill in Wayne, New Jersey. The spot was called The Anthony Wayne on Route 46. I was a young teen, and it provided me with pocket cash and a stage upon which to elevate my social standing among friends, strangers, and girls whose attention I craved. Never in my wildest dreams did I think this would evolve (devolve?) into a lifelong passion.

No, my dream was to go to college to study journalism and become a writer. Hell, I had already picked out a pen name: McCane. I read vociferously at the urging of my teachers, my mother, and my grandmother, and my thirst for understanding things and people through the written word never died.

Wait, who am I kidding? I had already won the hearts of the prettiest cheerleaders with my sensitive poetry and ability to weave witticisms into every paragraph. I felt sure my approach to love and fame was bulletproof. I had no doubt my word prowess would lead to multiple Pulitzers, a string of bestsellers, and an entourage of beautiful women.

The key was to earn a degree in journalism from the institution of my choice—Ithaca College. Ithaca is nestled in the idyllic town of the same name in the Finger Lakes region of upstate New York. I was sure my dad would gladly pay the freight. What could go wrong?

"No fucking son of mine is going to become a pansy-ass writer," my father shot back. "You're going to study business. The health care business."

My father served in the air force during World War II. When he returned home from the war, he and my mother married and produced five offspring in rapid succession over a period of just seven years. My father also teamed up with his buddies from the service to launch a packaging business when plastics first became the rage on account of their versatility. After they successfully designed and built packaging equipment, their company grew to an impressive size before my father was ousted in an early version of a hostile takeover.

Out of a job with five kids at home, he fell back on his field of study in college: pharmacology. He worked tirelessly to provide for us at a steep cost to the health and stability of our family. My father was never one to mince words or refuse a cocktail. His booming voice and lingering anger, seemingly over his ouster from the packaging business he had cofounded, led to bouts of excessive drinking. His wrath was inescapable and as hard as his liquor. When his eyes lost focus and blood filled his cheeks, his belt would come off. I never understood the cause of his rage, but I did feel the results.

During my junior and senior years in high school, I was an all-star basketball player. I made just about every all-conference and all-county team, and I even earned an honorable mention on the all-state team. In one game during my senior year, I scored thirty-four points in thirty-two minutes. But when I got home, I got no pats on the back or praise from my father.

"You could have scored thirty-six points if you hadn't missed those two foul shots," he snickered.

I reconciled our relationship, and we found our peace in his later years after my mom passed away, and he quit drinking as suddenly as Forrest Gump stopped running. He passed away in 1997. Despite his faults, he was my hero and the most influential person in the early part of my life. His example of hard work, service to others, and entrepreneurial spirit left their mark.

My father's stinging words about my dream of becoming a writer and his sermons about me never being good enough or smart enough rang in my ears. I am both tormented by and grateful for his words: tormented by their psychic wounds and grateful for the determination they sparked in me to prove him wrong.

Still, that was that. His pronouncement ruled. Writing would have to wait. Four years after matriculating at Ithaca College, I earned a bachelor of science degree in health care administration, cum laude, class of 1975. The piece of paper was as useless as a fifth prong on a fork. But I leveraged that into a master's degree in hotel and restaurant management from Cornell University. Armed with all that knowledge and credentialing, I found myself back to slinging burgers for a living.

Yet, as I look back, this pursuit has filled every inch of my body with food and wine sensations, knowledge, and experiences, both positive and negative, as it drained and filled my wallet. It also took me from New York to Washington, DC, to Chicago, Houston, and, finally, to Dallas.

In addition to burgers, I've created and sold tacos, pad thai, lasagna, chicken fried steak, vindaloo chicken, barbeque, and rib eye steaks. To wash it down, I've peddled Long Island iced teas, Genesee Cream Ales, Old Style beer, frozen margaritas, sake, mango lassis, ruinously expensive chardonnay, Champagne, and twenty-year-old Tawny port.

I've conceptualized better airport fare, upscale fast food, white tablecloth dining, family-style Italian cuisine, hot dog stands, and delis. I've traveled from Bangkok to Bombay (Mumbai); New York to New Mexico; London to Laguna Beach; Vancouver to Venice; Cabo San Lucas to the Cayman Islands; and Hong Kong to Honolulu. Over those years, I have dated and written poems to beauty queens, prom queens, and once even a drag queen! (It was a blind date set up by a friend who never suspected her friend was a man in drag. I won't share how I found out....)

This book is part travel log, part human interest story, and part real-life fairy tale seasoned with a healthy dash of saucy exposé. The message that my parents drilled into me as a boy lives on in this book: Be honest, don't blow smoke up anyone's ass just to appease them, don't compromise your values, and do not shy away from hard work.

The more succinct message of restaurant concept development virtuoso Phil Romano also lives within these pages.

"If you don't like it, go fuck yourself."

And so it begins.

Prologue

All I could see through the thin slit in the gauze bandages wrapping my head were shadows, outlines of people moving around me. Coming out of an anesthesia fog is disorienting, a feeling underscored by nausea, a headache, and dizziness. I surmised it must be nighttime. There was a sense of calm among the hospital staff. Their movements lacked urgency—a stark contrast to the typical hospital environment, where organized chaos rules, and where patients, guests, doctors, nurses, technicians, and cleaning staff shuttle equipment, patients, and treatments from here to there at a frenetic pace.

It wasn't until a day later that I realized I was in a special ward: the neuro ICU, or the neuroscience intensive care unit. A nurse was sitting dutifully by my bedside, occupying the middle slot of three eight-hour shifts. One bed; one nurse. How many other beds were in this unit? I had no idea.

I heard a steady beeping, which was, in a way, calming. My arm movements alerted the nurse that I was awake, trying to orient my surroundings. I could feel myself coming out of the fog in moment-by-moment increments. As I tried to move, I was assured by a voice that I was okay. I was instructed to not move so much.

My head felt like a sack of rocks. I had no idea that there were ten pounds of ice packed into the gauze mass encapsulating my head—cold, dead weight. Where was I? What was I doing in this place? That anesthesia and its concomitant fog were doing their job, even if they were gradually releasing their grip.

Then, suddenly, I realized I couldn't breathe. I tried gulping in air. No dice. I clawed at my mummified head, attempting to remove whatever was obstructing my breathing path. I could get no air flow through my nose or throat.

"What is it? Why are you struggling?" asked someone off to my left.

An alarm sounded. More shadowy figures, more voices, and intensified movement followed. Someone urgently began peeling the gauze from my face.

"What is it? Why are you shaking so much?" a man asked, looking intently into my now exposed eyes.

I moved my arm near my throat and mouth. "Are you having trouble breathing? Is that it?" the man asked.

I nodded as best I could. What I didn't realize was that, even though I was strapped down, the entire bed was shaking violently. After my surgery, I measured six foot four and 295 pounds, down from six foot six and 320 pounds at the time of my admission. The removal of a brain tumor from my pituitary gland didn't diminish my stature much—nor my ability to quake hospital furnishings.

No one could get me to stop shaking or clear my breathing passages. I could see a doctor grasping a pair of surgical scissors above me, but still, nothing made sense. All I knew was that I couldn't breathe. My panic escalated.

"I cannot help you until you stop shaking!" the doctor screamed.

Somehow, through all my disorientation and fear, I could finally comprehend what he was saying. I summoned help from the only place I knew I could go.

"Mom, if you're out there somewhere, I need your help," I called out through my breathlessness and the gauze. It was more like a whisper.

Still not breathing, still moving uncontrollably, my entire body suddenly went still as those words left my lips. The doctor immediately opened my mouth and inserted the scissors to cut and remove a dangling wad of gauze from my throat. Did this procedure take a minute? Ten seconds? I had no idea. But a sense of calm returned after he pulled the mucous-encased mass from my mouth.

I could breathe again. I hungrily engorged my lungs. The movements of those around me subsided. My mom, long ago taken from me in a freak auto accident, once again proved that she would always be there for me.

1

And So It Begins

It was a miserably cold night in late January 1991. After growing up in the northeast and settling in the south, I can confirm the truth of the cliché that the only thing between Texas and the North Pole is a barbed wire fence. When it gets cold in Texas, it's bone-chilling—times two.

I was newly separated from my first wife and healing from an operation to remove a brain tumor (benign). I had also sold my first business for peanuts and lost virtually everything I owned. And after all that, I found myself driving on Interstate 45 from Houston to Dallas to begin a new job as a manager-in-training at Romano's Macaroni Grill. Mac Grill was an upscale casual Italian restaurant with theatrical touches such as jugs of red wine on the tables and operatic singers strolling through the dining room.

I was in a borrowed red Dodge Avenger, long before cell phones and satellite radios were commonplace. I had $110 in my pocket, a hundred of which was a gift from a dear friend who knew how desperately I needed it. I was cold, lonely, and on my way to a city that I knew nothing about to start a job at a company that employed thousands. Of those, I knew only one: my ex-sister-in-law Teri Scroggins.

I'd never worked for a large restaurant company before. I'd always had an entrepreneurial spirit running through my bones, and, if truth be told, I harbored a bit of disdain for chain restaurants. The drive was dark and isolating. I might as well have been at the edge of the world. I was heading to the home of a family friend, where I was to live until I got my feet underneath me. But even that did little to assuage my sense of dread.

I was thirty-five years old with a master's degree in hotel administration from Cornell University, the son of very successful parents, and the brother of four siblings who all were well schooled and in well-paying, career-driven positions. I was on my way to a strange city to become a $35,000-a-year assistant restaurant manager. As low points go, they don't drop much lower. Or so I thought.

Around midnight and about halfway to Dallas, the headlights on the Dodge suddenly went out. The car was still running, hurtling down the highway. It was a dark, moonlight-free night with a multitude of stars smeared across the sky like thick cobwebs. This was no way to drive.

There was little traffic, no gas stations, and no sign of a small town. Then, just as suddenly as the headlights went out, the car's engine went dead. I coasted down the highway, trying my best to keep the tires planted on the pavement through the oppressive darkness. Luckily, I had slowed down enough to easily coast onto what I assumed was the shoulder. I sat there in the gloom, wondering what to do. Nothing came to mind. A sense of desperation washed over me.

Now, I'm, by nature, a confident man. But I could not overcome my despair. I broke down on the side of the road just outside of Buffalo, Texas, tears streaking my cheeks, wondering what my life had become. It was the same sense of panic I experienced after my brain surgery six months before. I opened my eyes, stared at the sky through the windshield, and called out to my mother.

"Mom, if you're up there, I need your help again," I cried. "I'm not sure I can get through this without you."

I closed my eyes and was overcome with a sense of loneliness. When I dared to open them again, a sudden flash of light flared just above the horizon. It brightened the entire sky for a fleeting moment. Then suddenly, the headlights in my car came back on. They illuminated the shoulder and the road ahead. I immediately reached for the keys, still in the ignition, and attempted to fire up the engine. It started. I pumped the gas to be sure I heard what I thought I had heard. The response was strong. I pulled onto the road, pressed down on the accelerator, and resumed my trip to Dallas, two hours due north, holding my breath just about the entire way.

* * *

I'm not sure what that sudden flash meant. Despite a Catholic upbringing, I'm not religious or someone who believes in greater powers at work in the universe. I'm just a simple man who lives in the moment and dearly misses his mother.

My mother and I had a bond that ran deep. During my high school years, it was just her and me at home, as my older brothers and sister were either off at college or out of college and beginning their careers. My father was there, too, but he was an alcoholic. He spent his days working hard at our family pharmacy, but his nights were spent downing drink after drink until he became belligerent. Those years were very dark.

My mother would often leave during these episodes to join her mother, who lived some thirty minutes away, and I would head outside and shoot hoops in the driveway. Sometimes, we would scavenge around the house, trying to find the bottles my father had stashed away. We'd replace the vodka with water, or at least severely dilute the spirit, hoping he wouldn't notice. He mostly didn't.

She also took me to hospitals, nursing homes, and homes for underprivileged youth so that I could join her in her volunteer efforts. She was trying to teach me humility, I think. Though people close to our family told me how much I reminded them of my father, my inner strength and sense of right and wrong came from her. During the winter of my sophomore year in college, she was killed in a one-car accident. I suffered unbearable grief for months. I often wondered how I could go on.

Ask me how I explain these phenomena—in the hospital and during my midnight drive—and I will tell you it was my mother reaching out to help me. Does it need to be more complicated than that? She may have left this world too soon, but she lives on. Somehow. In the end, what matters is that I made it to Dallas, started my job on time, and began this journey.

* * *

After leaving the hospital following my brain surgery, I had an intense headache. My head was packed in ice most of the day, and I had limited movement. So I watched endless TV. I had nothing but time on my hands to think about things, yet I still had absolutely no idea who I was or what I wanted to do with my life.

My struggle to conquer that tumor exposed other maladies. My marriage to my wife, Nancy, had been falling apart for months prior to my surgery, and though she stood strongly by my side throughout the ordeal, no amount of surgery could repair the rift that developed between us post-op.

I met Nancy during a Halloween party at Nino's, an Italian restaurant founded by Vincent Mandola, Houston's godfather of Italian cuisine. I was the general manager of this deep Brooklyn-style dining spot, which was one of Houston's hottest restaurants. It was a dining scene where you could catch any number of celebrities and movers and shakers coming in and out of its doors on any night of the week.

Frequent guests included Billy Gibbons, the late Dusty Hill, and Frank Beard—ZZ Top—when they were at the height of their rock stardom. I wasn't a fan of their music. I much preferred R&B— Luther Vandross, Michael Jackson, and Evelyn "Champagne" King. But the show that hit every time those guys drove up was unmistakable.

Their flashy classic cars and their leggy, scantily clad entourage of gorgeous women belied just how humble and likable they were. Big tippers too. Houston in the mid-1980s was still Boomtown, and nowhere was that more evident than in the dining room of Nino's. In addition to ZZ Top, high-powered lawyers closing mega-deals, Houston Oilers football players, and wealthy oilmen crowded the tables nightly.

On this occasion, I was dressed up like the Invisible Man, the character from the H.G. Wells science fiction novel of the same name. I wore a dark suit and fedora, with white gauze and tape covering the entirety of my head. I had slits for eyes that were hidden by sunglasses. I never spoke or removed those sunglasses. I walked around the dining room, gesturing with my hands or nodding my head. That's when I noticed this woman with long dark hair and an electric smile sitting at a table in the corner. Her laugh lit up the room, and her huge dimples were hypnotic. She was sitting with a tall, attractive blonde woman, and I flirted with them as much as an Invisible Man could flirt without talking or making eye contact.

"Take off the sunglasses," they requested as I passed by. I waved them off.

After they finished their dinner, they cozied up to the bar, urging me to unwrap my head, and as the evening wound down, I complied. My entire goal was to get the phone number of that dark-haired beauty, Nancy. Success. I called her the next day, and we agreed to meet for dinner. Before long, we were dating regularly. She'd been previously married to a "Mr. Texas" bodybuilder, and his routines had rubbed off on her. Her disciplined eating regime and her detailed attention to her own appearance were intimidating.

I'm not sure how it all developed so quickly, but we were married within a year. Her family loved me; I loved her spirit and genuineness, and we had, what would seem to most (including me), a storybook marriage. How I managed to fuck that up is something that haunts me to this day. No doubt my brain condition and the surgery that followed had a lot to do with it. I was never the same afterward.

But there were other reasons. Food-related reasons. At one point during my stint at Nino's, the Mandola family had an opportunity to purchase an adjoining lot next door. For years, it had operated as a dive bar before it was shut down. They purchased the lot and the run-down one-level shack with the idea of turning it into an Italian grocery and sandwich shop called Vincent's.

But the menu was missing something sweet. I'd always dabbled in making cookies and had a pretty good chocolate chip cookie recipe. So I made a batch and brought it in for the staff to try. They loved them and immediately urged me to make a batch to sell at Vincent's.

They were an instant hit, selling out most days. Several months later, we got a call from *Houston City Magazine*. They were conferring on us "The Best Chocolate Chip Cookie in Houston" award for their Best of Houston issue. Visions of Famous Amos and Mrs. Fields danced in my head.

So I turned in my resignation at Nino's, leaving a well-paying job to open my own cookie shop. It wasn't that I was dissatisfied or bored with the position. I was just ready to move on to the next challenge. And a challenge it was. I got a loan from Tanglewood Bank, where my wife Nancy had some connections, and tossed in $30,000—my entire savings—to launch Chip off the Old Block.

Over the next year, I worked my ass off for almost nothing. I'd chosen a bad location, was undercapitalized, and had no clue how to sell. But one day, a kind gentleman came into the store and told me how much he loved my cookies. Turns out he owned two ice cream shops, and he was eager to sell my cookies at his locations.

But even with his business, I was still failing. I was forced to close but was able to sublease the space to a baker who was looking to expand her operation. Win-win. Me? I was out of a job, had blown my savings, and had no clue what to do next. Then, my ice cream friend approached me with an idea.

"Why don't you buy one of my stores and operate it?" he asked.

I had no money. But he suggested I pay him out of my profits over time. I agreed. I was now the owner of an ice cream parlor that featured my locally famous cookies. Not long after that, my brain tumor was discovered, and I was forced to sell. I was riding my first downward spiral.

My tumor was first detected while Nancy and I were attending the wedding of a mutual friend in Annapolis, Maryland. We were there with my friend, Bruce Nash, a doctor who had married my high school sweetheart, also named Nancy. The night before the wedding, we went bowling, and I was complaining about fatigue and pain in my joints. Bruce asked me a few questions. He mentioned that I seemed a bit taller and heavier than he remembered. He somehow connected my joint pain and tiredness to acromegaly.

Acromegaly, or "giant's" disease, is a condition whereby the pituitary gland produces too much growth hormone, usually the result of a benign tumor known as a pituitary adenoma. That extra dose of growth hormone gradually enlarges the skeleton—hands, feet, jaw, nose—long after normal growth has stopped. Untreated, the disease promises an early death. Bruce suggested I find an endocrinologist when I returned to Houston. He never said anything about a brain tumor.

I followed his advice when I got back home and made an appointment with a well-regarded endocrinologist. My wife Nancy was traveling, and I didn't give the visit much thought. I figured I'd get a prescription and be on my way. When the doctor walked into the exam room, we spoke generally for a few minutes about my health—nothing specific, just conversation.

"Do you want me to tell you why I think you're here?" he asked.

"Sure, doc, what do you think I have?" I never mentioned my conversation with Bruce or what he suspected I might have.

"Acromegaly," he said.

"Bingo!" I replied flippantly.

He paused. "Well, let's discuss how we go about treating a brain tumor and what your surgical options might be."

I felt like my chest had just been hit with a cinder block. He continued to explain my condition, but I'm not sure I heard a word. Eventually, I left his office and drove around aimlessly. My world was spinning. All I could think was that I was going to die. Months of testing, worrying, and planning how my in-laws could help followed. My brain tumor was the size of a walnut.

My options to remove the tumor? Saw the top of my skull off and go through the top of my brain. Or undergo a new procedure called transsphenoidal surgery, where the upper lip is peeled back, and the surgeon enters the brain and retrieves the tumor through the nose. Great choices, right?

I chose the nose way.

* * *

I was adrift without a plan, and though no one could accuse me of being a philanderer, there was nothing that seemed to click between Nancy and me—or in my life in general. I'd always seem to wander, losing interest in whatever I was doing to look for a new challenge, even if it was risky. I never really mastered anything because I'm not sure I have that gene. I have no interest in dedicating myself to perfecting a single discipline. Not sure if that's because I'm afraid to fail, or if I just lose interest in things easily. Nancy saw this. And it scared the shit out of her.

I found myself detached from just about everything and everyone around me—except for my perky sister-in-law Teri. Married to my wife's brother, Teri was from New Jersey, just like me. But unlike me, she was a ball of energy and one of the most positive people I have ever known. She knew how much I loved the restaurant business, and she was employed by one of the largest restaurant companies in the US: Chili's. Years later, Chili's would change its name to Brinker International after its founder, Norman Brinker, purchased the Chili's chain and began scooping up other restaurant companies.

She'd heard Chili's had just acquired a successful restaurant in San Antonio called Romano's Macaroni Grill, founded by restaurant concept creator Phil Romano. She knew a few people who were charged with expanding the operation and offered to help facilitate an interview for me.

The person I eventually interviewed with became my most important professional influence, and we developed a close personal friendship. His name is Rick Federico. He worked with Grady's American Grill, another Chili's acquisition. Rick was one of the only people at Chili's corporate with Italian heritage. The story goes that when Romano sold a part of his interest in Macaroni Grill to Chili's, he insisted that someone of Italian descent take charge of the expansion. In the early 1990s, Rick was given the reins of what would become the first of many of Norman Brinker's successful national expansions.

I drove to Dallas a day early so that I could be on time for my interview with Rick the following day. But the interview was postponed for a few hours. So I met a friend for a round of golf that morning. I kept my clubs in the trunk of whatever car I was driving so that I was always prepared for spontaneous rounds. My game took much longer than I had planned, and I didn't leave myself enough time to change out of my golf garb.

I hightailed it to the Chili's offices in my golf clothes, soft spike shoes and all, and made it just in time for the interview. When I walked into Rick's office, I must have looked unprofessional at best. But it turned out that Rick was an avid golfer as well, so my blunder was a plus. We connected immediately. My background managing Nino's didn't hurt either.

A week after that fateful interview, I received an offer letter. I was to begin work as a management trainee at the Romano's Macaroni Grill in Addison, Texas, just outside of Dallas—the first expansion location after the Chili's acquisition.

In late January 1991, after a year of not working post-surgery, I began my new career. I was part of a team of twenty management trainees forming the leadership corps that would, over the next few years, turn Macaroni Grill, or Mac Grill as we called it, into a national powerhouse.

I showed up to work at seven-thirty in the morning on a Monday in early February—a few days after my "miracle" ride to Dallas. A chilling wind blew as I approached the back dock of the restaurant. The only information I had was the name of the general manager and my schedule for that week.

I entered the back dock door and was met by a manager in chef's clothes who introduced himself as Barry. He was thin as a rail and spoke with a heavy New York accent. That Monday morning—and every Monday morning—Macaroni Grill received a huge food order. This location was enormously successful, achieving $100,000 a week in sales, serving dinner only. This level of weekly sales was among the highest in casual dining at the time. Boxes were lined up everywhere, waiting to be unpacked to feed that demand.

"Let's get this shit put away," Barry said.

No flowery welcomes. No orientation tours. Just a dispassionate command to unpack and stock deliveries. I took a minute to put on my apron and take it all in. I had been a general manager of a big-time Houston restaurant. But here I was, an assistant manager, making a little more than half of what I'd made there and relegated to unboxing food. I toyed with walking back to my car and bolting from the whole scene. But I didn't. It was one of the wisest decisions I ever made.

* * *

Over the next four years, I would become an important executive in a growing team and a close and trusted confidant of Rick Federico. And why not? We were the same age, we both loved golf, we both were restaurant rats, and we had a chemistry that seemed natural. He had a quick and genuine smile, was intensely sure of himself, and was excellent at patting backs and making people feel comfortable. He sensed that I had more to offer than simply running shifts at a chain restaurant and that I wasn't at all intimidated by the hard work and pressure associated with high-volume venues.

I was also a workaholic. I spent seven days a week training, offering to work more shifts than I was scheduled for. I guess I was trying to escape my failed marriage, and, since I was living in a city where I knew almost no one, I buried myself in my work. Macaroni Grill in Addison, Texas, became my home-away-from-home for seventy-five to eighty hours per week. Sure, I might play basketball at a local health club every now and again (I had played a few years of small college ball), but not much else.

For weeks, I didn't have a dime of expendable income to my name. But once I had some pocket change, I spent it on wings and beers for my fellow trainees at a local Hooters after work. We were all in the same boat, betting our future on a fledgling concept that we all thought we could make better. And we did. Big time.

In the early years, Rick Federico would come into the Addison restaurant (Chili's home office was just five miles away) and query a manager—often me because I was there all the time—about things like recipes and equipment. He was collecting intel as we ramped up for rapid expansion. I shared my opinions. The tricky part—as is always the case when sharing opinions about someone else's baby, that someone being larger-than-life Phil Romano—is that you must be careful about how you present those opinions and with whom you entrust them.

Rick knew my background, so, when he asked me what I thought of our house-made red sauce, I didn't mince words.

"It's essentially dead blood and sour milk on account of how we cook the veal bones and Romano cheese in large tilt kettles overnight," I said. "Abysmal stuff."

He listened. The recipe was ultimately changed, and after some tweaking, we went on to create a dynamite fresh marinara that completely outshined the culinary sludge that preceded it.

Phil Romano may not have liked it, and I imagine there were many heated conversations with him voicing his opposition to changing his family recipes. But Rick was now in command of the concept, and it would grow with his predictable leadership style, not with Phil's bombastic ways and his cherished family recipes. The menu had an awful calamari steak. Every bite was like chewing on rubber. It was gone. We served a tough grilled pork steak, one that was difficult to cook exactly right. That was eighty-sixed.

Phil didn't want to serve lunch and felt that Macaroni Grill should serve dinner only, just like the original outside of San Antonio. Rick decided to serve lunch at all subsequent Mac Grill locations. Phil insisted no Italian restaurant he created would serve lasagna. It was too clichéd. Soon, we were serving dozens of portions of classic-style pan lasagna every night. Despite the initial protests,

Phil soon lost both his power and his interest in fighting the changes after he realized Rick had solid backing from Norman Brinker.

I would go on to become the first Mac Grill opening unit director, a position that had me traveling the country, opening new locations in cities from Kendall, Florida, to Vancouver, British Columbia. In three years, I opened just short of thirty restaurants in more than twenty cities in ten states. I made quick trips back to Dallas to attend meetings and update the design team on suggested improvements we'd learned on the job. I'd also pet the dog, wash some clothes, and find a date for the night. Then, I'd hit the road again.

For three years, I did nothing but travel with an energetic team of trainers. For two solid weeks, a team of about twenty-five people—cooks, managers, servers, and host trainers, even a voice teacher to train the opera singers staged in Mac Grill dining rooms—would check into a local Marriott Residence Inn or Hampton Inn and get to working our tails off.

It was a crew of prima donnas, scallywags, serious professionals, and an occasional misfit. Each was given a free ticket that included a good stipend, a free hotel room, a stage on which to perform in multiple cities across the country, and a team of fellow travelers who were attractive, available, and pretty much carefree. We partied into the wee hours.

It was an endless trail of one-night stands, too much drinking, and bars that were open too late and took too much of our money—our own version of the rock-star life. We lived a 24-hour lifestyle, with half of the day dedicated to sweat and rigor and the other half devoted to rewarding ourselves with indulgences.

We were the envy of the industry. We made local print and broadcast news, handed out precious passes in the best bars and strip clubs of that specific city to our dry-run practice sessions, and generally lived a charmed, fast-paced life. Many of those trainers went on to become restaurateurs of some acclaim and success.

I fell in love at least once during every opening. But the absolute truth of the matter is that I never compromised the business or my position. Yes, there were bumps in the road. The occasional lost rental car. Or the casual fling that went awry.

I wasn't naïve about the activities of the young men and women on my team, nor of the attractions that would more than likely manifest as all-night parties and shack-ups. But what happened behind closed doors generally didn't affect our work, so it was not a concern of mine. Mostly.

Years later, one of my superstar trainers explained to me that the male trainers had an expression they joked about while on the road. Sex to them was different from being in a relationship. Their mantra was, "You won't lose a girlfriend; you'll just lose your turn." These were not people looking for long-term relationships. They were competing for conquests. But one night in Tulsa put an interesting twist on that formula.

It involved one of our vocal trainers. She was six feet tall, leggy, and blonde as the blondest Texan—easily a Miss Texas candidate. Smart, too; very religious, and a trained opera singer to boot. One morning, I received a call at the restaurant from the head of Macaroni Grill training.

"Mark, we have a problem. Can you come back to the hotel?" she asked.

I hopped into my car and made the short ride back to the hotel. There, in the lobby, sat four people: my head trainer, two male trainers, and the female vocal trainer. I was greeted with hanging heads, no eye contact, and very flat facial expressions as opposed to the energetic "curtains up" guise I was used to.

My head trainer explained that the opera singer had somehow managed to have sex on separate occasions with both male trainers on the same night. In their room. At different times. Without the other knowing about it.

You must understand these male trainers were roommates. The realization by these three of what took place generated some very raw emotional reactions.

To this day, I am not clear how she executed the deeds, but somehow, there were hurt feelings, a little humiliation, and palpable tension in that lobby. All three were sent home, and replacements were summoned.

Despite the assorted indulgences, what we accomplished as a team was remarkable. And remarkably memorable. As much as I earned Rick's admiration for my thoroughness and the trouble-free transitions I executed from opening to opening, the bond we formed on the links was every bit as important. I'm convinced these interactions between golf swings were crucial to my rise at Brinker and the future opportunities that came my way. I have always searched for win-win situations, and my four years with Mac Grill were one of the premier examples of that.

Yet I had grown tired of the relentless travel. I needed a different position as I became more well-known among the Brinker executives; one where I could use my creative abilities to their fullest. It was never my ambition to become a C-suite executive. I don't have the play-by-the-rules mindset for a role like that.

As Rick and I spoke more about what was next, he asked me if I thought I could work with Phil Romano on a new venture Brinker was funding. It was a home meal replacement concept that was equal parts grocery store, wine shop, bakery, butcher shop, and grill. But it was all parts Phil Romano's over-the-top personality.

* * *

It didn't have a name, but it did have an analog: Harry's in a Hurry in Atlanta, an upscale convenience store with gourmet foods and premium novelty items all squeezed into four thousand square feet. Harry's was similar to the mom-and-pop neighborhood stores I grew up with in suburban New Jersey. It was also similar to the Polish butcher shop my father's father, Butch, had opened in Bloomfield, New Jersey.

There, I stocked shelves, cut meat, and helped make kielbasa, the Polish pork sausage. But I drew the line at making head cheese: that concoction of pig's ears, hoofs, gelatin, and who knows what the fuck else. It makes me wretch to this day when I see it in stores. My grandparents lived upstairs from that store, and it served as their own personal pantry and meat market.

This setting and upbringing were not unlike Phil Romano's own childhood in upstate New York. For some reason, he asked Rick if he could speak with me about this new project. Rick acquiesced. He had another trainee in the hopper to take my place as Mac Grill barreled toward two hundred units. Rick got his replacement and recommended I work with Phil on this new unnamed project.

I felt important and thought it was time I made my own mark on Brinker International. Phil gathered a small, hand-picked team, and we holed up in our own little wing in one corner of the Brinker corporate offices, which, at the time, was staffed with three hundred-plus people. It was a chance for me to rise to the level of my mentor and oversee the next big thing at Brinker. I felt like I'd graduated from the girl next door to the prom queen overnight. I went from reading the reports produced by others to writing reports for others to read.

As a side benefit, I had daily interaction with who most of us in the office thought was the best-looking gal at Brinker: Jana Hardgrave. Being single and on the lookout for my next romantic relationship, I was always scanning the crowd to find someone who caught my eye. Jana did. She was about five foot four, with a slim, athletic body, wavy dirty blonde hair down near her waist, and a come-hither smile that was as addictive as a narcotic.

Jana passed out the long, multi-page printed reports cataloging sales from the previous day for all the restaurants in the company. It must have been a gargantuan undertaking to assemble these things, and we all took them for granted. This was years before laptops, email, and smartphones. We did everything on paper, and each report was hand-delivered to the offices of each executive. I had finally made it onto that list.

Jana's office was in accounting—far from mine—and other than her distributing the sales reports, it is unlikely we would have ever met. She would drop off the report, and we'd often engage in conversation. Since Mac Grill was a darling of the company, and our team got extra attention because of our great performance, it made me a bit more interesting.

I asked a fellow staff member about Jana's potential availability and discovered she had recently broken up with a guy I knew from my department. I felt there was chemistry and mutual attraction, so I eventually got up the nerve to ask her out. She said yes. We hit it off immediately and began dating exclusively within a few months.

Not long after, I passed the test with her very attached mother, and it wasn't long before we were living together. This was a full four years after my divorce, and I felt it was a good time to try this relationship thing again, as I'd overcome my sense of failure with Nancy. We lived together for a few years before I finally decided I was ready to try marriage again. I proposed to Jana on a flight to Vancouver (in first-class) on our way to an Alaskan cruise. We spent our first week as an engaged couple in a suite on a visually dramatic cruise that was both refreshing and distinctly memorable.

* * *

As I was leaving Macaroni Grill, the best way I can describe the experience is I went from being somebody with luggage full of respect from executives to having to carry luggage for someone whose only concept of respect was the title of that famous Aretha Franklin song. Meaning that Romano showed—or seemed to know—little of the actual stuff. I'm not sure if the difficulties started on day one, but they were in full force by week one. I was now part of what would be the Eatzi's Market and Bakery team. I bid farewell to all my traveling Mac Grill companions of the past four years.

The common thread during my Brinker years was my relationship with Romano during the Mac Grill buildup and the honor of having been chosen by him (as I was told) to be part of Eatzi's. The next four months of my life became a challenge unlike any I'd experienced before or since. Eatzi's was a very cool project, but the clash of personalities amongst the team made it a living hell that melted all that cool very quickly. As much as I loved working with Rick and the Mac Grill team, I had entered a short phase of my life that was as gruelingly unpleasant as any I had ever experienced before or since.

As the late Texas Senator Lloyd Bentsen once said to Vice President Dan Quayle during the 1988 vice presidential debate: "Senator, I served with Jack Kennedy, I knew Jack Kennedy, Jack Kennedy was a friend of mine. Senator, you're no Jack Kennedy."[1]

I felt the same about Phil Romano.

I worked for Rick Federico. Rick Federico is a friend of mine. Sir, you're no Rick Federico, was what went through my mind daily for about four months.

But regardless of my disappointment, the experience threw me in the middle of launching a hot restaurant concept with entrepreneurs of ruthless ambition. The lessons learned, however painful, taught me a whole lot about what *not* to do when the next opportunity arose. But first, a short chapter on the worst few months of my career.

---

[1] "Senator, You're No Jack Kennedy," NPR, May 23, 2006, https://www.npr.org/2006/05/23/5425248/senator-youre-no-jack-kennedy.

2

Eatzi's. The Shortest Job I Ever Had

Norman Brinker walked into my office regularly by mid-afternoon to check on the progress of the Eatzi's project. We had been working for three months on the conceptual phase—which basically translated to *traveling to as many similar places as we could and making mental, physical, and photographic notes of all that we saw and tasted*. And while Phil Romano was the de facto head of the development process, Mr. Brinker was never comfortable casually checking in with him in his corner office at Brinker headquarters. So he came to me.

Everyone was aware of Romano's volatility, and during the concept development phase of the projects he did for Brinker International, the executives steered clear of him. While I have a background in concept development and have never been the starched corporate type, I like to think that Mr. Brinker recognized that I was the best buffer between the creative and the financial and organizational processes.

Brinker might have been one of the country's most prolific and freewheeling restaurant think tanks, but they were still a public company and were beholden to stockholders. Romano could have given two shits about financial accountability since Brinker was footing the bill. He also wasn't adept at communicating what he was up to, so that fell on me. My nickname amongst my close Brinker colleagues was "Shakespeare" because of my compulsion for communicating project progress through interoffice memos. My reports were always sprinkled with personal observations that were far afield from typical technical reports.

"Mark, I'm going to Washington, DC, next week, and I was wondering if there are any places I should visit?" Mr. Brinker asked me one day. Washington was rich in on-trend, urban-style neighborhood markets and specialty grocers.

"I really think you ought to check out the Sutton Place Market in Old Town Alexandria and the new Dean & DeLuca's Market on M Street in Georgetown," I replied. In his characteristic understated style, Mr. Brinker nodded and thanked me before disappearing down the hallway.

Our cross-country research excursions ranged from small neighborhood grocery stores to one-hundred-thousand-square-foot supermarkets, from farmers' markets to food courts, and from classic delicatessens to old-style butcher shops. Stops included Balducci's and Zabar's in New York City, Oakville Grocery in Napa Valley, Marche in Toronto, Foodlife in Chicago, and Jamail's Grocery in Houston.

In the mid-1990s, the hot new restaurant segment was labeled *home meal replacement*. That segment emerged dozens of years before but was never labeled as such until Boston Chicken (renamed Boston Market) hit the scene. What Boston Chicken proved was that you could generate enormous sales without hosts, servers, bartenders, and highly trained kitchen staff. For example, the typical Chili's could generate $2.3 million in annual sales from six thousand, two hundred square feet of space.

But Boston Market could rack up $2 million on less than three thousand square feet—all without the associated costs of the typical full-service restaurant. In the restaurant biz, sales per square foot are the primary barometer of success. And Boston Market was far exceeding full-service restaurant averages.

Large restaurant companies like Brinker stumbled all over themselves, trying to replicate this budding segment. Brinker's answer was Eatzi's Market and Bakery. In many ways, this was a complete departure from Brinker's market strengths. But the company was determined to cash in on this emerging market—even if they didn't entirely understand it. Whether the move was "covering your bases," a desire to dominate, or maximizing stockholder value, the effort eventually went haywire after a short time, succumbing to the weight of complexity and costs.

As work proceeded, Mr. Brinker again came knocking on my door. As he approached my desk, his look was uncharacteristically pensive. Though he always had an overall even demeanor, on this day, he flashed a frown.

"I went to those places in Washington, DC, you suggested I visit," he said. "Did you know, Mark, that Dean & DeLuca has over twenty different kinds of goat cheese? And their butcher case has every cut of meat you could imagine. We're not doing something like that, are we?"

Backstory: I knew from the start of the project that Mr. Brinker was enamored with simplicity and had no interest in complex concepts. His vision was more aligned with another Dallas chain: 7-Eleven. But instead of processed convenience foods, it would be stocked with great take-home foods prepared in a commissary kitchen and delivered to each location. Picture a 7-Eleven with refrigerated cases filled with baked meatloaf, macaroni and cheese, freshly made soups, salads, casseroles, and light ethnic foods like lasagna. Also, a wall of bread loaves, cheeses, beverages, and a few frozen items.

But Romano had no use for anything easy and simple. Three thousand square feet could not contain his creative vision, nor could it express his enormous ego. His concepts overflow with complexity and stagecraft. His work with Brinker produced an Italian destination (Macaroni Grill) with split kitchens, opera singers strolling the floors, and dozens of fresh gladiolas under theater-style lighting. He also created a Mexican concept called Nachomamas, which Mr. Brinker renamed the less controversial Cozymel's. Many of Cozymel's moving parts were as complex as Romano's Macaroni Grill.

A storm of conflicting goals was about to collide.

Back to Mr. Brinker's visit to my office. It was all I could do to stammer my way to an answer that would neither be completely truthful nor totally false.

"To be honest, Mr. Brinker, I'm not sure we're at a point where we're ready to define specifics like that," I said. "We've reviewed a lot of concepts and possibilities, and it's still on the drawing board."

I swallowed hard with the half-lie, half-truth. When he left my office—though he appeared satisfied—I knew there was no way he was buying what I was selling. The lid was about to come off.

"What the fuck did you tell him?!" yelled Romano as he stormed into my office doorway almost immediately after Norman left.

Though not physically imposing, Romano was a natural-born, spring-loaded fighter. He could be as intimidating as a man twice his size and weight. Brinker executives kept their distance. Romano was a wildcard, and his explosiveness—though widely known— was anathema to Brinker culture. He even unnerved me, who stood nearly a half foot taller and fifty pounds heavier. After another dozen f-bombs and veiled threats, he leveled his final command.

"You don't tell him a fucking thing about what we're doing. Ever!"

If my hair could have been singed by words, I would have been bald on the spot.

It didn't take me long to realize this was not for me. As much as I loved the creative process and considered Eatzi's my ticket to financial success and high-profile industry recognition, there was no mistaking one critical detail: Each of my paychecks was signed by Norman Brinker, not Phil Romano. It wasn't my place to be the middleman between two strong but incredibly different personalities. And though I did not fear Romano, I had no desire to lock horns with him daily if not hourly. Life is just too damned short.

I took the rest of that afternoon off and headed to my house to consider my options. In the quiet of my home, I was better able to calculate my options. I didn't have many good ones, but one decision was certain: I decided to resign rather than get swept up in a gigantic power struggle. I called Lane Cardwell, the Brinker-appointed overseer of the project, and told him of my decision over the phone. I chose not to get specific about my personal and professional issues with Romano but shared enough so that my exit would be understood and communicated to Mr. Brinker.

An hour later, my phone rang. On the other end was a very different Phil Romano—quieter and more controlled. His concern? Not how we could work this out—you eventually learn that Romano doesn't so much want you as need you—but rather how my resignation could be spun so that it didn't reflect poorly on him and his pet project.

Despite his bluster, Romano desperately wanted to be accepted at Brinker, and if my resignation—after just four months—was anything but smooth, that acceptance might be compromised. I wasn't an inner member of the Brinker brain trust, but my opinions were valued. For my part, I had no interest in anything but moving as far away from the Eatzi's project and the mercurial Romano as possible.

My confidence was growing, and my reputation and résumé were both firmly cemented from four-plus great years with arguably the most revered restaurant company in the country. Some of my old swagger had returned, and instead of fearing what might be next, I welcomed the unknown.

Eatzi's eventually opened to wild success. Its eclectic mix of sights, sounds, aromas, and infectious energy had visitors smitten. It expanded to twenty units across the country and, in the end, was much more Romano's elaborate vision than Mr. Brinker's conception of a food-centric 7-Eleven. But eight years later, the flaws of the concept began bubbling to the surface: deficiencies that had been evident to me from the first month.

Eatzi's probably would have had more staying power—with success in mid-markets—had it expressed Brinker's vision rather than the flaming unwieldy meteorite Romano created. It inevitably burned out, bogged down by high labor, development, and operational costs combined with low unit sales as it expanded across the country.

Brinker eventually sold Eatzi's to an investment firm that took another five years to drive it into the ground. But before it completely imploded, Romano, along with a few partners, purchased it back, closing all but the Dallas flagship store while attempting to reanimate the concept by aligning it to Romano's original vision. Fueled by Romano's immense energy and willpower, they were able to keep it afloat by stirring in a little of Mr. Brinker's commissary vision, deploying a regional "hub and spoke" paradigm. It's been slowly expanding—albeit in the Dallas-Fort Worth area only—since, with six locations as of this writing.

Oddly, it's as impossible to not admire Romano as it is to respect him. It is that dichotomy that is his legacy as much as the concepts—both failed and successful—he's scattered across the restaurant landscape. If Romano had been less a bully and more of a collaborator, Eatzi's might never have flamed out.

The true success of a highly regarded restaurant company like Brinker relies on mutual respect and careful (sometimes too careful) collaboration among seasoned industry veterans. Romano's idea of collaboration was more "just do what I say and keep your opinions to yourself." Oil and water. Could it have worked? Yes, but that's like saying if zebras didn't have stripes, they'd be horses.

3

Sam's Café. Proving Ground

While I had been able to hang on at Brinker doing odd jobs, it turned out there was no landing strip for me in any of the concepts that Brinker had acquired or was expanding. I casually asked around, but there wasn't a fit as natural as Mac Grill had been. It was becoming clear that I wasn't really one of the Brinker insiders, despite my qualifications and availability. I realized it was time to leave.

One of Norman Brinker's greatest strengths was the development of a tree of associates who would use his company as a platform for others to earn credibility and a springboard to launch their own successful companies. People moved on. Chris Sullivan and Bob Basham, two founders of Outback Steakhouse, were Brinker alums. Rick Federico worked with Norman for years at Macaroni Grill and went on to become chairman of the board at the wildly successful P.F. Chang's and Pei Wei Asian Diner. Creed Ford, one of the original Chili's executives, left to grow a concept called Johnny Carino's. He also expanded Phil Romano's Rudy's BBQ. Hal Smith, the former CEO of Brinker, left to start his own successful restaurant empire in Oklahoma City revolving around a restaurant called Charleston's.

That exodus was leaving the hallowed halls of Brinker thin on leadership and talent. With so many restaurateurs taking leave to start new companies, those who remained—almost universally a uniform crew of ex-Chili's guys—were left to run the dozen existing or newly acquired concepts. That's when I noticed the first crack in the Brinker foundational cement that would eventually bring the company to its knees a few years after Norman's passing in 2009.

Consumers began recognizing that Brinker brands didn't address their expanding and diverse dining habits or cultures. It wasn't uncommon for three concepts—Chili's, Grady's American Grill, and Macaroni Grill—to sign leases or buy property right next to each other. The foods and appeal of these concepts stopped reflecting the audiences they were trying to attract. So it's no surprise my talents and expertise were no longer aligned with the company. I was anything but cookie-cutter; I was a Tommy Bahama shirt in a sea of Brooks Brothers button-down collars.

And finding a restaurant executive job, particularly with a résumé that included an impressive stint at Brinker, was not all that difficult. Smaller, fledgling upstart companies were forming and operating throughout Dallas-Fort Worth area. I was soon interviewing with several. These upstarts fit my mold better, and I was prepared to merge my recent corporate learnings with the dynamic entrepreneurial atmosphere that permeated Dallas in the mid-1990s.

* * *

I accepted one of the first jobs offered to me by Jack Baum, a dynamo with a striking charismatic presence. In late 1995, Jack and his right-hand man, Bruce Lazarus, had begun expanding a concept called Sam's Café, with two locations in Dallas and a pair in Phoenix. A bit younger than I was, Jack was an East Coast guy like me and could convince you—if he put his mind to it—to walk on hot coals without fear of getting burnt. He passionately shared a vision for Sam's Café that convinced you it would be the next Chili's.

Yet Sam's Café was not Jack Baum's brainchild. It was started by Academy Award-winning actress Mariel Hemmingway (*Lipstick, Woody Allen's Manhattan, Star 80*) and a few local investors (Mariel's nickname was Sam) who couldn't develop it. Jack swooped in to buy it. He had a few other successful operations, including a seafood restaurant in Dallas's historic West End district called Newport's. Jack sold me hard on developing Sam's into a national chain not unlike what was in the Brinker stable.

To Jack, Bruce, and their small board of directors, I was the missing link in their plan. Our interview was short, sweet, and, as we would find out years later, mutually beneficial. I took the director of operations position at a comfortable salary and rolled up my sleeves. With only four restaurants serving southwestern cuisine—gaining in popularity at the time—there was a solid base to work from. And for Jack—being the salesman he was—raising expansion capital was not an issue.

It was a good formula. The Sam's Café team was diverse, and Jack was an appreciative leader who engaged in the business without meddling. Bruce was a genius with numbers. I had a team that included two chefs: Dallas-based Travis Henderson and Phoenix-based Tudie Johnson—both had been with Jack for about a decade. My team also included two Phoenix-based area directors: Kevin Hale—a handsome John F. Kennedy look-alike—and Mary Warder, a stunning brunette who eventually became a close friend. Beyond that, we had young and inexperienced managers, some office staff, and a full calendar of restaurants to build using the resources Jack had secured.

My first goal was to familiarize myself with the food and get to know the people. I immediately hit it off with Travis, and he taught me the ins and outs of the operation. While I am not a trained chef and have never attended a cooking school, I have worked in many kitchens. Travis immediately took me into the Sam's Café kitchen at Preston Center, put an apron on me, and started at the beginning.

"Do you know why you roll a lemon?" he asked.

"No, Chef. Why do you roll a lemon?" I replied, my words dripping with sarcasm.

It doesn't take long to understand many of the basics in this business. Rolling a lemon is done to release the juice from the segments inside the skin so that when you cut into it, you have ready-made lemon juice, which can then be used in sauces and other recipes. He smiled at my wisecrack, and immediately, I knew my East Coast cynicism and his Texas cowboy charm would work just fine together.

Travis and I became inseparable, not only during my Sam's Café years but for many years afterward. During my three warp-speed years with Sam's Café, we traveled, drank, and skirted serious trouble from coast to coast. With Tudie and Travis at the culinary helm, we created food that was unique but accessible. We made pasta with chipotle sauce, black beans and cotija cheese garnished with cilantro, adobo chicken sandwiches with blistered poblano peppers and avocado mayo, and more—food that was different but recognizable.

We owned the edgy, approachable foods category, and markets responded favorably. We executed with precision, a byproduct of my Brinker training. I recruited young and energetic people for my team who had a vision for growth that mirrored my four years at Mac Grill. We strived to create an adventurous, chef-driven menu and a hip southwestern ranch vibe. It worked!

Harris Design out of Dallas created a color palate and décor that was unlike anything existing in the US. They incorporated heavy wood timbers, sunset colors, lots of leather, and hints of adobe living. It was a gorgeous setting. But we soon realized that the name "Sam's Café" didn't really capture the flavors and the vibe. The board had long urged a name change, and we quickly settled on "Canyon Café," with a new logo and a design overhaul.

With a new look and feel, we soon rivaled other well-known brands. We'd managed to catapult southwestern cuisine onto the national scene. We paired these new tastes with a beverage menu that explored tequilas at a time when this spirit was largely ignored. Our prices were moderate, our service was expertly orchestrated (servers wore denim shirts and bolo ties), and the music was energetic.

We earned a National Restaurant Association "Hot Concept" award, which was given to just a handful of dining creations every year. We opened about four or five restaurants annually over the next three years. From Atlanta to Kansas City to Seattle, we commanded prime real estate. We were able to recruit top-level management and chef positions, and our culture was the envy of the industry.

You might go to a Macaroni Grill if you wanted opera and endless wine; a Houston's if you wanted starched white shirts and impeccably consistent food and service; or a Cheesecake Factory if you wanted sensory overload and huge plates of food. But you chose Canyon Café if you wanted a new culinary adventure along with an escape to ranch country with a completely new vibe. And it worked no matter where we opened.

With strong per-unit sales, we were soon courted by larger companies that wanted to absorb our revenue streams, culture, and team. It wasn't long before Jack found a solid partner for our continued growth: Apple South, a franchisee of Applebee's Neighborhood Grill + Bar restaurants and Chili's most aggressive competitor. The same year they bought Canyon Café, they scooped up five other casual dining brands, including Don Pablo's Mexican Kitchen, McCormick and Schmick's, Hops Restaurant Bar and Brewery, and a few other smaller chains.

Almost overnight, we were a part of a huge restaurant conglomerate. It was a whirlwind romance that made Jack and his investors overnight multimillionaires. We all celebrated with parties and big bonus checks. It was about as good as it gets—for a while, at least.

* * *

I've never fully understood why people want to fuck with successful formulas, but I've come to understand that it's probably just human nature to not leave well enough alone. It was clear that as we became part of Apple South, they really didn't understand Canyon Café and our business. Our team at Canyon was young, aggressive, and unburdened by structure.

I'd brought a semblance of organization from my Brinker years (along with a great handful of managers!), but our spirit and culture were decidedly looser and more entrepreneurial. We knew how to have fun, often into the wee morning hours. We were hardly cut from the super conservative cloth at Apple South. We didn't so much lock horns, but we just continued to go about our business, doing what attracted them to us in the first place.

Them loving us was one thing. Them knowing how to nourish that love was a whole other proposition. The Apple South team thought differently than we did, and battle lines eventually began to form. On one side was our team of energetic, knowledgeable, independent, and headstrong restaurant professionals who felt Canyon Café didn't need fixing. On the other were the starched, earnestly inside-the-box corporate minds at Apple South who saddled our team with Doug Czufin, an unlikable leader with a thin culinary background. From the jump, Czufin and I had no chemistry.

He tried to wedge his way into our talented and energetic posse, calculating that his position would earn him immediate credibility. What he didn't realize was that we needed nothing from him. He had nothing to add. He was a chef but couldn't measure up to what we already had in Tudie and Travis. He had no understanding of southwestern cuisine and the nuances that made it so unique.

Instead of taking the time to understand and endear himself to us by rolling up his sleeves and participating, he chose to play politics and wield influence with his bosses at Apple South. His suggestions and ideas were inane. Czufin insisted we reduce kitchen labor by having some of our signature items produced at a commissary kitchen, despite the fact that we weren't yet big enough to outsource our food preparation. Our entire team rebelled.

There was arrogance on both sides, I admit. Our attitude was, "Just shut the fuck up and give us the money to continue doing what we do best." Czufin's thinking was, *Let me show you all the things you're doing wrong and how to fix them whether you like it or not.*

Hilton Eades, a friend from my Mac Grill years, had an expression anytime someone tried to fix things that weren't broke.

"It's time for the blind to lead sighted!" he would blurt in his thick southern drawl.

Czufin was our blind guy. He and I locked horns frequently over his decisions that took hold with the support of the suits at Apple South. Our entire team bristled. Jack was caught in the middle, in the uncomfortable position of playing peacemaker or siding with Apple South. Things came to a head one late Friday afternoon in Dallas.

Jack called me into his office and asked me to close the door. I sat down across from him, sensing his discomfort, yet I had no clue why. He fidgeted as I stared at him curiously. Jack and I were close but not outside of work. It was a relationship built on mutual respect, and we'd settled into a great rhythm. Within minutes, he told me I was fired.

It wasn't contentious. Jack and I had drifted apart but not to an uncomfortable degree. The decision made no sense to me, but I was tired of fighting with Czufin. I'd started to hate him and all that he was trying to do. If firing me meant he finally won, then so be it. I accepted the decision quickly and quietly. Jack and I hugged, talked about what a great run we'd had, and said our goodbyes.

I'd grown to like Jack. A lot. And I still do to this day. As Michael Corleone said often in *The Godfather*, "It's not personal. It's strictly business."

I called a couple of close friends in the company—Travis and Mary Warder—to let them know what happened. Mary and Andy came over to Jana's and my house, and they showed up dressed in all black. We all had a great laugh over that! Then, my phone blew up. Jack must have immediately called several other key people after I departed the office because the news spread quickly.

This team that I had helped build and mold, a team that had zero turnover and a whole lot more money in their pockets, started to reach out to express their loyalty. Mary, Andy, Jana, and I drank generously, while the phone kept ringing. Czufin may have won the battle, but the war was far from over.

One of those phone calls was from Jack Baum, who rang me a little before midnight. He asked if we could meet, suggesting we both go back to the office. I got in my car, drove the two miles to the office, and met him in the dark, eerily quiet space. He shared that he'd received an enormous amount of negative feedback over my firing. The pressure was enough for him to reinstate me within hours of my termination. It was surreal. It was the only job I've ever been fired from, and my firing lasted just six hours.

I'd known that Czufin was probably behind my firing, but I was over it and ready to move on. I'd known all along that Czufin was mostly an ass-kisser at the Apple South offices and that he'd never been able to penetrate our team's sense of unity. The strength of our team proved more powerful than his feeble power moves. I accepted the re-hiring that night and drove back home to sleep it all off.

I came back, but, truthfully, the damage was irreparable. We'd reached a kind of stalemate in our growth trajectory. We now had a new team guiding our development and finite resources to grow Canyon Café and our new sister concepts. All these new siblings, particularly McCormick and Schmick's and Don Pablo's, along with Canyon, cost major dollars to build. We were now just one of the concepts in Apple South's pipeline, and our growth stalled to a near stop.

On top of that, just about everything we did was intensely scrutinized. Czufin was doing his best to consolidate his grip on Canyon Café as my influence waned. My patience wore thin. I was tired of fighting. I resigned.

This time, there was no mourning or fanfare. The brilliance that was Canyon Café was snuffed out by a group that never understood what we were about and proved they were incapable of assimilating our culture into their corporate DNA. We'd been a darling in their eyes, enough so that they paid a huge price to purchase us. It made no sense, but there was no turning back—for any of us.

We'd done what we set out to do: we made the board and Jack very wealthy, turning a tiny fledgling concept into a national player—not a bad three years of work. I'd met some great people, made a bit of a windfall, and have no regrets. Jack and I kept in touch, but it never was the same. Eventually, he left the company.

Chef Travis Henderson left to open his own restaurant with one of our superstar general managers, Amy Burgess, as a partner. They achieved celebrity status in a polished Dallas steakhouse called The Place at Perry's. It was truly the best of times.

Until it wasn't.

Travis was tragically shot and killed by a Dallas SWAT officer in 2012 after he stepped out of his vehicle brandishing a handgun. Witnesses surmised he was bent on self-harm—suicide by cop. The years of travel, some well-known dalliances, and an overdose of the good life took their toll. The deep sadness of losing Travis was felt by all of us who'd traveled and worked with him. It was a crude exclamation point concluding the Canyon chapter.

Many of my manager friends also left, and only a few from our original team remained. One of those who stayed even as Canyon Café was collapsing was Pete Botonis, a bright young manager I hired in Phoenix. Remember that name. It will come up many times in subsequent chapters.

Apple South's love affair with their mega-restaurant dream eventually came crashing down. The company divested all its newly acquired restaurants within a few years of purchasing them. What was left of Canyon Café was picked up by Doug Czufin and Bruce Lazarus. It wasn't long before they put Canyon Café out of its misery. In email exchanges between us, Czufin blamed me for Canyon Café's demise. I would remind him that the reins he so desperately coveted were in his hands. He cavalierly brushed off my criticisms.

Jana and I took a year off to enjoy our moderate-but-comfortable windfall from the Canyon Café sale and traveled to Italy to drink wine, eat pasta, and stroll the streets of Rome. It was well-earned, sorely needed, and replenishing. I came back ready for my next adventure.

I dusted off my Tin Star menu and called Morgan Hull. I was sure we could take Dallas by storm with this exciting new approach to great food. The lightbulb over my head was high-wattage and ready to shine.

4

Tin Star. Great Concept, Awful Partnership

Long before I left Brinker International, I had spent time noodling with potential restaurant concepts. My work with Eatzi's reinforced my belief that I could do my own thing and create a successful concept from scratch. Over the years spent opening new restaurants, I had met quite a few super-talented people, from chefs and managers to financial analysts and designers. I traveled all over and had seen many great restaurants in different parts of the country, so my understanding of trends was broad.

But for some reason that I don't quite understand, my focus was on fast-casual dining: an emerging and enormously popular trend in the mid-1990s. Fast-casual is a hybrid of full-service quality cuisine melded with the convenience of fast food—but at slightly higher prices and in much nicer atmospheres. My initial attraction to fast-casual was that it was pretty much an everyday experience. We eat every day, but we only dine occasionally. I wanted to be—and saw more opportunity in—everyday eating.

Few of these concepts were hitting the press at the time apart from two in Texas: Cafe Express in Houston, developed by Lonnie Schiller and Chef Robert Del Grande, and La Madeleine in Dallas, launched by restaurateur and chef Patrick Esquerre. Both featured chef-driven cuisine with counter service at lower-than-fine-dining prices in accommodating environs. Plus, they had speed-of-service that no full-service restaurant could match.

My first foray into this budding segment was a concept I called Tin Star with the tagline "Salsa, Smoke & Sizzle." I was enamored with the flavors of the southwest, a love that was bolstered by my recent experience with Canyon Café. But I was beginning to realize Asian flavors were on the rise as well. My friend Rick Federico was killing it with the heavily Chinese-influenced P.F. Chang's, and I just knew this was the tip of the iceberg in the realm of Asian flavors.

Tin Star was going to merge these flavors, and I needed a chef with the ability to finesse this dance. I found that chef in Mac Grill alumnus Morgan Hull—a quiet genius of Texas flavors and as serious a family man as you'll ever find. He wasn't a hulking type, but I always felt small in his presence, even though I was taller and weighed as much as he did.

He had piercing eyes and a sense of humor drier than the Sahara. We were not kindred spirits—I was a Yankee, and he was Texas-born—but we complemented each other as well as anyone I've ever worked with, other than Travis Henderson from my Canyon years. I admired his dedication to the study of food, was awed by his culinary knowledge, and never tasted anything he prepared that didn't become "the best of" what I'd ever eaten in that genre.

I enlisted one of Morgan's college friends: graphic artist Amber Brown. She was every bit of Texas that Morgan was and was easy on the eyes with blonde hair and an energetic aura that enveloped her presence. She immediately understood my vision, and the three of us put the original menu together in no time flat. But a funny thing happened on the way to bringing this new concept to life: Sam's Café, aka Canyon Café.

I had walked away from my Canyon Café chapter with around $400,000—not a killing but a decent sum for my three years of work. I wasn't complaining. I'd worked tirelessly, traveling endlessly to open new restaurants across the country, and gave 100 percent of myself to the cause. That sum financed Jana's and my Italy vacation, and we spared no expense on this excursion.

Returning refreshed from my time off, with some cash in the bank and new motivation, I began assembling a team for this new venture and rang Chef Morgan Hull. I set up a limited partnership, engaged my friend Ira Tobolowsky to compose the legal entity, and established my first LLC. Along with family friend Dave Buck, I invested enough to have a 51 percent stake in the Tin Star partnership.

Raising the balance of the capital is always easier when potential investors with much deeper pockets know that the creative director has skin in the game. I not only had skin, but I was the majority partner, so the balance of the money needed was easy to gather. Ira had insisted that I be the majority owner in the event of disputes.

But an entrepreneur rarely thinks about what can go wrong—only about what can be created and the rewards that can come with it! I recruited the rest of my investors, worked with Morgan on the menu, and created a plan to bring Tin Star to life. In my quest to complete the partnership, I reached out to Rich Hicks, a manager I had worked with at Macaroni Grill. We'd opened a new location together in Little Rock, Arkansas, and another in Plano, Texas. He was a sharp kid, a go-getter, and a guy who was quick on his feet.

His family was close friends with Rick Federico, my old Mac Grill colleague. In fact, it was Rick who suggested I track down Rich as he was looking for his next gig and might even have a few bucks to put into the venture. What I needed most was more money, and Rich, his dad, and his dad's friends soon became our limited partners.

Limited partners typically have no real power other than access to reporting information, a distribution of any profits, and bragging rights to their friends over their restaurant ownership status. It also allowed the risk to be more thinly spread out and is often the preferred structure for start-ups like Tin Star.

With the help of a great commercial broker, we were able to secure a space in a new building under construction in the Uptown area of Dallas, just north of downtown. Vincent and Robert Carrozza, the owners of the building, loved the Tin Star concept and firmly believed in my vision for fast-casual in their new office building. We quickly inked a deal on a corner space. It was small, but Morgan and I designed an efficient space that aligned with our menu and our vision for Tin Star.

The concept was counter service, and our menu was focused on great tacos (in 1999, this wasn't yet a trend, so we felt we were leading the curve), some salads, and a few platters of food that made it an attractive dinner option. Our goal? To build and expand Tin Star into a national concept, piggybacking on what I'd accomplished with Canyon Café. My confidence was strong, and the partnership was primed and ready. Restaurant chains were still expanding at a breakneck pace, and we wanted in!

Rich and I got along great. In the time it would take to build the restaurant, we spent pretty much all our free time together because neither of us had other work. Tin Star was our singular focus. We shared a love of golf, and with the proceeds from my Canyon Café windfall, I joined a private golf club in Dallas to tune my game. Rich was an excellent player, and we pushed each other to raise our level of play during this lull.

Our relationship was highly productive. We were entirely in sync with what we felt we could do with Tin Star. Time passed quickly, and once Tin Star opened, it came out of the blocks with the usual bumps and bruises associated with a new business. But we all worked overtime to make the necessary tweaks to our vision and quickly earned a great review and cover photo in the *Dallas Morning News* weekly *Guide* insert that propelled us to warp speed.

In 1999, well before Yelp and other online review platforms, the only way people found out about new restaurants was through reviews in newspapers, magazines, and by word of mouth. I drove up to our tiny place on that Friday night after that glowing *Morning News* review, and the line to get in was around the block. Our tiny, forty-seat restaurant was packed. It was pure havoc but the good kind. From that day forward, Tin Star would never be a sleepy, little startup.

* * *

The Tin Star business continued to explode. We generated national attention and garnered a coveted spot on the cover of one of our industry's nationally distributed magazines, *Restaurant Business*, in January 1999. The magazine tabbed Morgan and me for the cover shot of their issue that highlighted "Concepts of Tomorrow," and we shot into the national conversation.

The beginnings of our modest growth plans were firmly in place. Our investors were happy. Our landlord was ecstatic. Morgan's food was being lauded in just about every publication. We were developing an excellent reputation in the uber-competitive Dallas dining scene. I'd purchased a vintage 1957 Thunderbird convertible and had license plates personalized with "Tn Star." I parked it right in front of the restaurant every day.

It became a Tin Star landmark, and all day long, people would take photos of themselves (long before Instagram) in front of it, helping to create a cult-like following. I let parents put their kids in the driver's seat for some great photos, and generally, we became a stalwart in the Uptown community. We brought on a few more good friends to help manage the growing business and started planning a second location in north Dallas.

Morgan and I had settled into a routine, with Morgan opening the restaurant every day (we served breakfast tacos), and me coming in just before lunch and working through dinner. Rich filled in the cracks, and as I spent more time on store number two, Rich became the more visible face of Tin Star.

At first, Rich did extremely well, and his ability to schmooze guests was a strength. He was comfortable in the dining room—though never in the kitchen—and eventually voiced his desire to have a more substantial role in the company. Morgan and I had a very clear vision for the menu and concept direction, as we'd done 100 percent of the menu iteration. Along for the ride as a minority investor, Rich wanted more attention as well as a bigger say in menu development and the expansion strategy.

It became an issue as our reputation grew. Our meetings became contentious. For example, Rich wanted to remove the skin from the chicken breast on our fried chicken salad. Morgan and I thought the crispy, crackly skin was an essential flavor and textural component. Everyone else was doing a skinless fried chicken salad, and ours was intentionally different and more crave-able. It was emblematic of our edgier food focus, so Morgan and I wouldn't budge on the issue.

We suggested to Rich that he bow out of the food discussions during our meetings, a veritable "get the fuck off of our lawn" declaration. But Rich would have none of it. He continued to push the menu in a healthier direction, mimicking recipes from Chili's, a concept he was very familiar with from his Brinker days. Morgan and I saw this as dumbing down the concept, and we put our foot down.

Rich started acting more like a majority owner than a manager with a small investment, and our differences became stark and combative. It was apparent to Rich that he wasn't going to get his way. He became sullen and petulant. As we continued to build the business,

Rich was disconnecting from it. Tin Star was gaining strength and growing, but Rich wasn't growing with it.

He was handsome and single, and Uptown was a hotbed for the young and restless and all that came with that. I'd come into the restaurant to find him flirting with guests and not really paying attention to the business. I'd finally had enough.

I told Morgan that I was going to meet with Rich. At that meeting, I fired him from the management team. I couldn't do anything about his partner status, though I did offer to buy him out. He was surprised by my decision and wanted to debate it, using his minority share as leverage. But I was under no obligation to retain him as a manager, and he had no recourse but to stand down.

Unfortunately, that wasn't the end of it. Since Rich had brought several investors to the table, including his stepfather, it wasn't long before they voiced their opposition to my decision. And though they were a minority group, they were essential to the growth of Tin Star with their deep pockets. Rich's stepfather and his executive buddies came into the restaurant to persuade me to reverse my decision. It was a true Mexican standoff.

There would be no second location without resolution of this issue. And to his dad's way of thinking, that resolution had to include Rich in a major role. Non-negotiable to me. The damage had been done. Rich was out. Period. It was clear to me that he had no place on our team, and through his arrogance, he'd burned the bridge to both me and Morgan.

During this episode of animus, an anonymous visitor walked into our little corner of Uptown. He sat alone at the bar during midafternoon when I had taken a break and gone home. When I returned, Morgan handed me a very plain business card, telling me that this gentleman had stopped by and that he wanted to let me know. At the time, I thought nothing of it. The name meant nothing to me and didn't seem to have an affiliation with any restaurant company. I stuffed the card into my pocket and forgot about it.

The following day, I got a call from Rick Federico. He was now the president of P.F. Chang's China Bistro, a beast of a concept out of Phoenix that was expanding rapidly. I was surprised to hear from him.

"Do you know who visited Tin Star yesterday?" he asked.

"Yes," I replied. "Paul Fleming. He left me a business card. Who is he?"

Paul Fleming is the P.F. of P.F. Chang's. He had partnered with Philip Chiang of Mandarin Restaurant fame in San Francisco to create P.F. Chang's China Bistro, merging their names and their combined vision. Their restaurant was the hottest concept on the national dining scene and was expanding rapidly. It was super cool, and everyone loved it. At one time, it was the busiest restaurant in Dallas.

Rick went on to tell me that Paul was thoroughly impressed with Tin Star and wanted to talk to me about how I'd created it and how I might help them do something similar for P.F. Chang's.

"Can you fly out to meet with Paul next week?" Rick asked.

"You buying the ticket?" I joked.

* * *

One week later, I was sitting across the table from Paul Fleming over lunch at a Houston's Restaurant in Phoenix in the Biltmore area. Turned out it was more of an interview than a conversation. On the flight home, it occurred to me that I had possibly discovered the solution to my Tin Star dilemma.

A day later, Rick called and told me my interview with Paul had gone well. I'd passed the test, and they wanted to make me an offer to join them in Phoenix. Paul had determined—after that interview and his single visit to Tin Star—that I was the missing link to a new

MARK H. BREZINSKI

concept he was concocting, that I might be the guy to help make it happen.

I'd never planned to do anything but focus on growing Tin Star. But under the circumstances, that wasn't going to happen. It wasn't an easy decision. Tin Star was on the cusp of greatness. Once Rich was removed as a manager, the team gelled nicely. But the reality was that we didn't have the money to grow without those minority investors led by Rich's stepdad.

Finding new investors wasn't an option. Too complicated. It had all been a murky mess until the offer from Rick was factored in. I had to make a tough choice, but that was made easier by my wife Jana. She was born and raised in Texas and had long wanted to get the hell out of Dallas and its high-powered bustle. She was very excited by the prospect of moving to Phoenix. On our frequent visits there during the Canyon Café years, she had developed a spiritual connection to the desert. That helped tip the scales. After some deliberation, I reached out to Rich's stepfather and requested a meeting. I offered to step away from Tin Star and sell my ownership to the group for face value.

As it happened, Rich had not found another job after I'd fired him, so he was available to take over the company. It didn't take long for Rich's dad and their minority group to make their decision. Rich would be back in, and I was out. Tin Star was still getting lots of positive attention, and the opportunity for growth was wide open. After some haggling, I collected $250,000 for my Tin Star interest. I was determined to become a free agent.

The ink wasn't yet dry on my sale to the partners when I accepted a position in Phoenix to work on this as-yet-unnamed new fast-casual concept. It was less than a year after we'd opened the first Tin Star, and here I was, driving to Phoenix to begin another phase of my career. I'd been handed a bit of a lifeline by Rick and Paul, and I was ready for the challenge of starting over yet again.

As I was making my move to that city, Rich quickly took charge of Tin Star, and from my conversations with Morgan, I knew he wasn't long for the ride. Rich was now free to do whatever he wanted. He even changed the tagline from "Salsa, Smoke & Sizzle" to "Smoke, Salsa and Sizzle." He was determined to erase my influence and firmly implant his own. I'm sure that in my absence, I took all the blame for everything that might have been wrong and that Rich was determined to fix Tin Star. C'est la vie, right?

Months later, I read an article where Rich claimed he was the founder and took credit for creating the concept. My instinct was to lash out à la Czufin, but frankly, I was just glad to get my original investment back so that I could start fresh in a new city. Looking back, it was as much a win-win as I could have hoped for.

* * *

Postscript: Tin Star, under Rich's leadership, went on to grow successfully as a franchising company. Rich had always liked the idea of collecting franchise fees, and I imagine his partners must have agreed. After one additional company store, they franchised exclusively, growing it to about twenty units.

But Tin Star hit a brick wall when the booming success we experienced in Dallas didn't translate to markets outside the DFW area. I wasn't surprised. Rich simply didn't have the leadership skills or, with Morgan gone, the culinary sense to propel the concept to higher levels. He also lacked real estate experience, wasn't versed in training and franchisee support, and didn't have a strong work ethic.

Tin Star ultimately stopped growing and tipped over into a slow decline. After several franchisees began closing their locations, just a shell of a company remained. With not enough franchise fee income and no real company-owned store revenue to count on, Tin Star had become a "Concept of Yesterday" instead of a promising national competitor. In an ironic twist, years after I left, Rich put the company up for sale, and an investor package was submitted to me for consideration. I passed. What began with great promise—earning a "Concept of Tomorrow" designation—became just another restaurant failure dotting the landscape. Damn shame.

But I had little time to think about the past and what might have been because I was on a wild culinary ride. My work with Paul Fleming and Pei Wei's Asian Diner was up next. And boy, did I love that episode. Fortunately, it loved me too.

5

Pei Wei Asian Diner. Solid Platinum

It was a beautiful spring afternoon in Dallas in early 2005. I'd been with Pei Wei Asian Diner for four years, riding the high of an eye-catching expansion plan that was exploding before us. Life was good. Real good. My wife Jana and I were home in our modest ranch-style house in north Dallas, enjoying the freshness of the day.

Suddenly, the doorbell rang. At our door was a FedEx deliveryman, holding just a slim overnight envelope: the kind that usually contains a smaller envelope within. I signed for it, closed the door, and ripped open the tear strip. The contents of that envelope changed our lives. Forever.

* * *

Paul Fleming is as far from a Phil Romano-like China shop bull as you could ever imagine. Quiet, of slight build, and introspective, he had not a bad word to say (at least in public) about anyone. He was humble and unpretentious. Paul could have afforded a Bentley or a Porsche, a closet full of designer clothing, and a spacious walled-in gated mansion.

Instead, he opted for oxford and Tommy Bahama shirts, khakis and jeans, loafers without socks—always without socks—and a well-used pickup truck. He lived in a modest house on a golf course that anyone could get a glimpse of if they hit a bad slice or, if you were a lefty, a duck hook.

My new assignment? Help industry darling P.F. Chang's China Bistro and creative founder Paul Fleming develop a "small box" version of their legendary Americanized Chinese restaurant. At the time I came aboard, the idea was just a collection of papers in a cardboard box in the corner of Paul's Scottsdale office.

That box included original design work, name options, and a few menu mock-ups seemingly created by an experienced conceptual development team. This box represented hundreds of thousands of invested dollars (probably more), yet nothing happened to stick. Amid their incredible growth spurt, my guess is that no one or group of people within the company had the time to dedicate to this new project. All resources—financial and manpower—were clearly dedicated to the enormous task of growing P.F. Chang's.

Paul was not exactly left alone in the pursuit of this "small box" idea. With my deep experience in the fast-casual realm, I was offered up by Rick Federico as Paul's de facto right-hand man. By the time I landed in Phoenix in 1999, P.F. Chang's had grown into a thirty-five-unit national chain generating spectacular returns. They were setting a new standard in restaurant performance rivaled by only two other national chains: The Cheesecake Factory and Houston's. Crazy. Stupid. Money.

An initial public offering catapulted the company into the stratosphere, making all the executives millionaires several times over. The P.F. Chang's story is a dramatic lesson in entrepreneurialism gone wonderfully wild. To be invited to participate on the fringes of this success was an honor and a privilege. Despite that success, the brain trust at Chang's clearly saw a niche to expand their restaurant empire by branching out into a more casual, accessible, and less costly model that dovetailed their great reputation.

When Paul had visited Tin Star months earlier, he determined that I could help them do just that. So there I was in my small conference room office in Scottsdale, with a cardboard box on my desk and a pile of sample menus, graphics, floor plans, and studies conducted by a consulting firm. The conclusion of these studies was that the dining-out market was ripe for a fast-casual Asian concept. Making sense of all this information, which included handwritten notes and freehand line drawings, was a daunting task.

Not to mention that the P.F. Chang's inner circle had no idea who I was or why I was there. During one meeting, Chef Paul Muller, an accomplished culinarian who had been with Chang's for years, introduced himself.

"Who are you?" he asked.

"I'm part of the Pei Wei team," I answered. He tossed me a blank look.

"What are you doing with Pei Wei?" he asked.

"I guess I'm in charge of bringing it to life," I answered.

That blank look transformed into bafflement. My loosely conveyed assignment was to put some flesh on the bones of this new concept. I was to make sure it resembled Chang's but was not identical; to introduce new menu items but not too many; to serve alcohol but not too much; to make sure menu prices were less than Chang's but not too cheap; to utilize some members of Chang's support staff but not too many.

I worked in Paul's very conservatively furnished suite of offices in Fashion Square—a few desks, a conference room, and simple décor and furnishings. It easily could have been a small office for an insurance company. And conveniently, it was a stone's throw from the original P.F. Chang's in the high-end Fashion Square Mall across the parking lot. For weeks, I listened to and absorbed input from

Paul, Rick, and just about anyone else who had opinions on this small-box concept.

I had at my disposal any of P.F. Chang's kitchens, the talents of Chef Paul Muller and food consultant Barbara Tropp, P.F. Chang's CFO Bert Vivian (another Brinker alumnus), and a powerhouse real estate department led by fellow Cornell alum John Middleton. There was also Shannan Metcalf, an enormously attractive, tall brunette who would become my assistant and coconspirator.

We spent hours accumulating competitive information and doing our best to assemble this massive jigsaw puzzle called "Pei Wei's Asian Diner." We held meetings, traveled to analyze other concepts, and operated on a loose timeline.

"Let's fly to San Diego," Paul would say. "There's a concept I want you to see."

The next thing I knew, we were aboard his private jet, taxiing down the runway for a day trip to the West Coast. Our driver whisked us from the airport, and for about four hours, we ate noodles, dumplings, and just about every kind of Asian food we could find from mostly mom-and-pop restaurants. It was a blast, and that one-on-one time with Paul was invaluable to me.

His passion for creating something authentic yet unmistakably Chang's-inspired was crystal clear. I was flattered he invested his trust in me so quickly. On a whim, he'd come into our office.

"Let's go do some Asian food shopping and cook some dishes," he said.

Soon, we'd be off grocery shopping before planting ourselves in Chang's kitchen for hours on end, cooking, tasting, and sharing flavor ideas. At the time, I knew practically nothing about Asian cuisine. I couldn't whip up a simple order of fried rice if my life depended on it. But over those first six months, I was privileged to cook side-by-side with some of the best, most accomplished Asian chefs in the country: Paul Muller, Barbara Tropp, and, on one occasion, Philip Chiang himself.

I learned the difference between oyster sauce and soy sauce; sweet chili sauce and sambal; rice noodles and traditional Chinese egg noodles; oil-velveting and stock-velveting, and so on. It was a whirlwind learning experience, and my arms and wrists have the scars to prove it. I was building the perfect background to create the next big thing in Asian restaurants, right?

Still, the picture was not coming into focus.

Opinions on design, menus, alcoholic beverage service, and pricing varied considerably. I was the new kid on the block, so I'd listen and defer. But I was getting a clear picture of why it had taken so long to move this project in a productive direction. No one had taken the reins, and the culture at P.F. Chang's was so inclusive that just about everyone had a say.

Yet Paul was growing impatient. I never asked him, but I sensed he was growing tired of this democratic process of concept development. We had a bad case of creative constipation. I felt I should take charge and spearhead this project.

* * *

After some game-planning with Shannan, I decided to approach Rick directly.

"Paul really liked my efforts at Tin Star, right?" I asked. "So why don't I just apply what I learned there to Pei Wei's Asian Diner and see what I come up with?"

Rick nodded. But it wasn't that simple. I was no neophyte at this creative thing. I had essentially created Tin Star with Morgan Hull in about the time it took this team to fill a cardboard box with ideas and market research. But due to my deferential nature, I wanted to do this work with full consideration for everyone involved. But that's not possible, and deep down, I knew it. Creativity is never a democratic process. Something had to give.

I decided to attack my assignment from a different angle. I whittled down Pei Wei's menu from sixty items to thirty, composing a handwritten menu I thought would work. It incorporated many of P.F. Chang's most popular dishes but added some of the food and flavors Paul and I had discussed during our many conversations. It was concise and straightforward, with strong and diverse Asian influences.

I began to visualize Pei Wei as a "best of" menu, absorbing flavors from the far east Asian continent while keeping ties to the dishes P.F. Chang's had popularized nationally. It was well received by everyone. Well, almost everyone. Paul was hesitant because this direction diverted further from his original vision than he was prepared to accept.

As any restaurateur will tell you, the menu is a very personal thing. If you fuck with someone's vision for the food—especially when you don't have their level of experience or success—you're skating on thin ice. After he registered the positive response from the rest of the team to my presentation, his relationship with me noticeably cooled. I continued to work in his offices, but our paths crossed with less frequency. Our lunches together dwindled. The private jet flights stopped.

I sought Rick's advice. He advised I let time pass, and Paul would come around. And he did. Maybe he started to like the approach. Maybe he liked me. Or perhaps he liked coming into the office and seeing Shannan's ever-present gleaming smile. Maybe he was onto his next thing, which inevitably would come about. His appearance and low-key demeanor belied his insane drive and energy. Paul preferred to stay away from controversy. He channeled his commentary through Rick.

But it was clear to all of us that Pei Wei Asian Diner had taken a turn. My focus now was on refining the menu that had gotten such a positive response. The box I had been working from, the one that contained an assortment of dead ends, was emerging in three dimensions. The kitchen was taking shape, and the design and layout were coming into view.

After Paul's eventual blessing on the project, the heat turned up. My role in the new P.F. Chang's division was firmly cemented, and decisions started flowing through me. Pricing? Shannan was on it. Food sourcing? Chef Paul Muller handled it. Colors and surfaces? Our brilliant designer, Brian Stubstad, took charge. Site selection? John Middleton had developers and landlords lined up.

Once everyone knows the direction, the best teams rise to the top. I had earned the respect of everyone involved by being more assertive, and I was finally clear on what we could create. But I needed one final piece. I went head-hunting at my old employer, Canyon Café, for a general manager. It took some convincing and a few nasty emails with my old nemesis, Doug Czufin, but I finally landed that key component: the wildly charismatic Pete Botonis. He accepted a cut in his pay and my offer of a chance to get in on the ground floor of an opportunity that comes around maybe once in a lifetime. We never looked back.

After six months of project development, the first Pei Wei Asian Diner opened south of Phoenix in the suburb of Chandler. We struggled at first to find our rhythm, as almost all new ventures do. But we gained momentum after a few months and enjoyed booming sales. We had projected a moderate $30,000 a week in sales to be conservative. But we were soon topping $60,000 and sometimes $70,000 a week!

Our well-publicized affiliation with P.F. Chang's brought us business from all over the Valley. Executives from other restaurant companies across the country flooded our dining room as everyone wanted to see what P.F. Chang's had come up with. Our small, fifty-seat dining room was often a "who's who" of top-level restaurant executives. They marveled at the activity and long lines.

We were serving between one thousand, two hundred and one thousand, four hundred people a day, busting at the seams of that tiny restaurant. We anxiously anticipated our first critical review. It was published in the *Arizona Republic* on December 14, 2000.

It was abysmal.

"Stay a-Wei," read the headline. "Pei Wei Eastern-Inspired Menu Goes South."

A bevy of quotes from the article hit me like Texas hail on a metal roof.

"The food here simply isn't very good."

"At times this place is like a human salt lick."

"Dreadful potstickers should be cited for impersonating a dumpling."

"Didn't the executives test these recipes before inflicting them on an unsuspecting public?"

"The best thing about Pei Wei is that it's cheap food. That's cheap food, not good food."

My heart sank. I composed a three-page response to reviewer Howard Seftel and his editor. In it, I accused the notoriously anti-chain Seftel of being vindictive and unprofessional. His comments were biased and lacked accuracy, but the editor said there would be no further comment—they backed their reviewer's right to his opinion.

I was livid and a bit concerned about Paul's and Rick's responses. Immediately, that concern was put to rest. Rick called me and shared his viewpoint about Seftel and his previous reviews. He surmised that the article wouldn't hurt us, that our loyalists would continue to frequent Pei Wei and drown out the memory of that awful review.

He was right. Our sales increased.

The mood in the corporate offices was upbeat, and the attention we received was off the charts. Paul would send me late-night emails—all of which I've kept to this day—warning me to not get used to these high volumes. I responded that I saw no reason sales would dip if we operated like we were capable of doing.

Sales never did drop. At worst, they evened out, and our growth was front and center at P.F. Chang's board meetings. A second site was quickly approved in Arcadia, Arizona, and the booming sales were replicated. Rick pulled me aside with a request.

"Mark, based on the early success of Pei Wei, the board of directors has approved a more rapid expansion, and the decision has been made to branch out into a second market," he said. "We've chosen Dallas. Do you want to stay here in the Valley and direct growth in Arizona, or do you want to move back to Dallas?"

I hesitated. I asked for a day or two to think about it. But the answer was easy. If I stayed in Phoenix, Pei Wei would forever be the brainchild of the Chang's team. I'd experienced how often people came into the restaurant asking for Paul or Rick. I didn't mind, but I also wanted to have more autonomy and to not always have to live in their shadows. I felt I had earned that.

But it wasn't only about me. It was about telling Jana about the offer. She had settled in well in Phoenix, and we loved our home and our new group of friends. Yet she was not the type to draw a line in the sand, and she supported whatever decision I thought was best. Dallas it was. It took about three months to solidify the deal, sell our house, and move back.

When you're at the helm of a new concept that strikes gold, as Pei Wei did in 2001, the entire industry takes notice. Landlords, developers, and an army of managers start lining up to become a part of that success. I went on to open more than thirty restaurants in seven states, including Oklahoma, Kansas, and Arkansas. In five short years, Pete and I developed our Pei Wei market area into the largest in the country.

Our growth strategy became the model for the company due to strong sales, great profitability, and an unassailable brand reputation. In many ways, we were the envy of the fast-casual industry and many of those in the P.F. Chang's restaurant family. Pei Wei racked up awards on top of awards and was referenced in industry media outlets as the "platinum standard" of fast-casual concepts. Sales volumes often rivaled those of the smaller market P.F. Chang's restaurants.

A Pei Wei opening in Lubbock, Texas, nearly topped $90,000 a week. It was an incredible high! It was a time of free-spending, with steak dinners, lots of entertaining friends and family, and the sharing of success with everyone who had helped us achieve it—and even some that hadn't! Sometimes, we'd be at a restaurant enjoying dinner and pick another table at random in the dining room who looked like they were having fun. We'd pay their check without them knowing it. It was hard work, hard fun, and a no-holds-barred adventure.

* * *

Which brings me back to that FedEx envelope. Inside was a check for $1.3 million. Being one of the early founders of Pei Wei, I had been granted a block of shares in Pei Wei Asian Diner. It was privately owned stock divvied up amongst a team of ten people who were part of the original work. With each successful opening and the skyrocketing growth of the company, the value of that stock climbed—that is, until it was purchased by our parent company, P.F. Chang's China Bistro. That check represented my share of the buyout. I was now part of the larger community of P.F. Chang's.

I immediately went on a rush to upgrade everything. I came home one day soon after that FedEx delivery and told Jana that I had found our dream house. I whisked her into the car for the five-minute drive to Dallas's prestigious Preston Hollow neighborhood. The house I selected was about three times the size of the home we had purchased just about two years earlier.

We didn't need that house. Jana and I had no children, nor had we plans to start a family. So a house that size wouldn't be something we could ever fill, at least with a family.

But I wanted that house. This may have been foolish, but I was insistent. Jana was far more conservative than me and was always comfortable with less.

But I'd also done my homework. The homes in that neighborhood had increased in value by 1.5 percent per year for the past dozen years. I convinced Jana that this was not a frivolous purchase but an investment. It was not all that difficult to see my logic and get caught up in the momentum of it all. Jana agreed, and we did the deal.

Our new home was six thousand, three hundred square feet and located on the dead end of a gorgeous tree-lined street; one of many other large homes. There was a small pond with a fountain in the middle of the neighborhood. We hired a top-tier interior designer and purchased more than $250,000 worth of new furniture and décor to supplement our existing furnishings. Our haul included Persian area rugs, some noteworthy art pieces, one-of-a-kind sculptures, and a kitchen decked out with every appliance in the known universe. I made plans to build a second-level deck with a spiral staircase winding down to a new patio with an outdoor fireplace.

"I promise, this is the last home we will ever live in," I told Jana.

The best part? We still had hundreds of thousands of dollars left and more to come with the growth of our Pei Wei market based out of Dallas.

It was a dream home in every sense of the word. We held parties, cookouts, and Christmas gatherings. We splurged on every event. It never grew old, and my income continued to rise. Though my stock had been purchased, my position as a Pei Wei market partner continued to provide more than enough resources to support whatever we wanted to do.

We purchased a getaway condominium in Scottsdale. We helped friends with loans and generously doled out money to friends and family with surprises like cars and expensive watches. We took semi-annual trips to New York City to see Broadway shows, always sitting center orchestra, while staying in luxurious Manhattan hotel suites. We even toyed with buying an apartment in New York's Upper West Side.

But I never stopped working, nor did I grow complacent. I simply knew how to enjoy the fruits of my labor, and the team supporting me was the best I've ever had. I justified my actions by telling people that I didn't have children, but I felt like the patriarch of the Pei Wei family. It wasn't pride or self-satisfaction that began to erode my position. It was boredom.

Pei Wei stopped evolving. I grew irritated with what I was witnessing as we expanded. The cutting-edge, aggressively versatile menu and design started to take a back seat to monstrous growth. Chefs Mark Miller and Paul Muller and I continued to meet monthly to create new dishes and flavors for the menu, but our work was never incorporated.

Pei Wei was now firmly entrenched in the financial arm of the public company, and it thirsted almost exclusively for growth and financial gain. I grew disenchanted, regardless of how much my bank account smiled back at me.

"Don't be a fucking idiot. Look how fat we're getting!" each statement would scream at me.

I'd never made this much money before, and I still haven't to this day. But the lack of creative spirit and the stagnation of the concept exhausted me. I told Jana I wanted to resign. By this point, she was convinced I was bullet-proof, so she supported my decision. Looking back, it was an incredibly selfish move on my part, one that impacted the two of us. It was the beginning of an unfortunate turn in our relationship.

* * *

I departed Pei Wei after a whirlwind and successful seven-year run. All thriving concepts like Pei Wei have a chance to rapidly expand and create value for developers. If you're first out of the chute with a top-notch team, others will invariably try to unseat you. But the task is Herculean.

Several concepts—Pick Up Stix and Mama Fu's, to name two—attempted the feat. But none succeeded. We had earned national awards, were consistently rated at the top of the industry in national guest surveys, and enjoyed a reputation as the preeminent fast-casual Asian concept in the country.

"Are you sure you know what you are doing?" Rick asked me when I submitted my resignation. "You're leaving millions on the table. For what?"

"I'm simply tired of the transformation from creator to administrator," I answered. "I need a new challenge."

"Mark, you don't have to leave. Think it over, please," he replied.

But I'd made up my mind, even as I knew I was taking a huge risk. And that, as they say, was that. Pete took over my Pei Wei market area and went on to grow the business for many years to come. I was happy to see him succeed. It was really all I could have hoped for when I recruited him with the promise of great opportunity.

I have never suffered the spirit-deadening tendency of corporate types to take what makes a concept special and turn it into a vehicle to simply make more money. Anyone who knows me well knows that. And while I love to spend cash, money alone is hardly what drives me. Rolling up my sleeves and working hard with my teams only to come home to an email inbox overflowing with administrative work numbed my excitement.

Pei Wei had lost its edge. It had become too vanilla for my taste. We should have developed a more expansive menu to cover a broader range of Asian cuisines. Years before I left, John Middleton and I had proposed we develop a smaller, more versatile prototype with a focus on takeaway business to squeeze into dense urban locations.

I had also suggested we take a hard look at some of our market partners who weren't performing and consolidate them with those who were. But ultimately, I stopped making suggestions because my influence dramatically waned. I didn't want to go down with the ship.

In my parting email to Rick, I warned him of the track to stagnation that Pei Wei was following. It took about a half dozen years, but eventually, my prediction came true. Pei Wei became just another concept flailing away to stay relevant in the face of declining sales and a fractured leadership.

A private equity firm invested in the company, and things only got worse as the focus on the financials grew more intense. My distance from the concept I helped create became more pronounced. I have no tolerance for looking back, cataloging "I told you so's," or stoking cravings for sour grapes. Ultimately, amid my high six-figure yearly income and seven-figure bank account, it all boiled down to a single question: what's next?

As it turns out, what was next would bring me to my knees and unleash hell in my life. That next was called Bengal Coast.

6

Bengal Coast. My Dream Come True

My wife Jana and I checked into the Taj Mahal Palace hotel at the Gateway of India in the heart of Mumbai after a week in Bangkok. Our time in Thailand was energetic and eye-opening. Bangkok is a very sophisticated and modern city for such a small country. What we observed on the ride in from Mumbai Airport to our hotel, however—about twelve miles that took an hour in our private car—was far different from what we'd seen in Bangkok.

It was our first time in India, and whatever notions we had about what we were about to see quickly vanished as we traveled the main roads from the airport to the hotel. Most roads were not paved and had no recognizable traffic signs or sense of order. They were jammed with bicycles, motorbikes, and very small cars. Many of the cars had people hanging out of the windows or onto the bumpers, comically weighing down the entire vehicle as if it were nothing but a simple frame overstuffed with humans hitching a ride.

The hour-long trip was dominated by car horns, near crashes, and a ride as rough as a dirt bike track back in the US. We were not in Bangkok or Tokyo anymore. I often referred to this circumnavigation of the world tour (Japan to Bangkok to Mumbai to London and, finally, to Dallas) as a confirmation trip, not an inspiration trip. Because I had already made up my mind that my next restaurant would serve upscale, modern Indian cuisine—*the P.F. Chang's of Indian cuisine.*

The difference? I would do it with a smaller partnership and would retain a 51 percent share of the venture, building it with a smaller team while exerting far more influence over the outcome. Mumbai is the most modern of Indian cities, and I was hoping the inspiration I drew from its urban culture would align with what I was imagining at the time.

I invited celebrated chef of southwest cuisine Mark Miller (Coyote Café, Raku, Red Sage, and author of more than fifteen cookbooks) to join Jana and me on this Asian tour. I paid him $25,000 to hang with us for two to three weeks as we tested food—sometimes five or six meals a day. (Jana often bowed out after the first two.) Miller mentored me in the flavors, cooking styles, and nuances of what we ate. He was my culinary guide. We also had a tour guide to ensure we hit the best places while steering us out of harm's way.

We checked into the opulent Taj hotel and made it to our assigned rooms. I'm no room snob, but I'm larger than the average person and have a self-diagnosed case of claustrophobia. I don't like sharing a bathroom, even though my wife was as gorgeous naked as anyone you may have ever seen in any feature film. (Nicole Kidman in *Eyes Wide Shut* comes to mind.) I simply need lots of room, and the room we were assigned just wasn't that room. Armed with a large cache of traveler's checks and a platinum American Express card, I went back to the front desk and asked if we might secure an upgrade.

"Of course. Let me show you some rooms," the hotel manager replied.

Off we went on a tour of the top-floor suites of this legendary hotel. We picked out our favorite: a two-bedroom suite with a large dining and living area (something like 1,500 square feet). The bell staff took all our luggage from one room to the new room quickly and silently, and we settled in.

The new suite was about $1,500 a night and came with a personal butler who was like a ghost. I'd leave the room for a quick morning walk and return to fresh juice, a paper, a buttery pastry, and a menu from which to order more food. There were always fresh flowers, and that politest of butlers disappeared even when you thought you were looking right at him.

His tip at the end of our stay was probably equal to his weekly salary—compliments of an attitude determined to spend our entire $100,000 travel budget over twenty days, or $5,000 per day. It doesn't seem like much today, but in 2007, it was plenty. We stayed at the hotel for a full week and explored every street stand and applicable restaurant, rooftop bar, and club.

* * *

Mumbai, along the bayfront, teems with high-rises and fancy clubs. But venture just one block off that high-rise row, and you'll have visions of *Slumdog Millionaire* greeting you as you look away from the sparkle of Chowpatty Beach. The walk from the hotel to the center of the city made Jana cry. We passed dozens of the poor and destitute who lived on the street. Young children tugged at our pant legs, begging for anything we could give them.

It was a daily occurrence during our week there, and even with the protection of a well-paid personal guide, we were always mindful of how far off the beaten path we wandered. While we could easily afford the luxuries of the best of Mumbai, neither Jana nor I were raised with this kind of wealth. It filled us with uneasiness to recognize the wretched living conditions of so many people while we spent the vast fruits of our labor exploring the city. It was difficult to wash off at the end of the day from our hands and our memories.

It wasn't sixteen months after our visit that Pakistani terrorists stormed that same Taj hotel, killing thirty-one people over four days of siege, burning it from the top down. I remember looking at the images on TV, shaking my head, knowing that those rooms on the top floor were where we stayed. It was an odd thing to watch, and all I could think of was how many of the friendly and gracious hotel staff—including our personal butler—had most likely lost their lives in that brutal attack.

During our stay, we never experienced anything more threatening than street beggars, usually little children who worked for larger networks of thieves. We were anxious to make our way to London, a town I knew well and a city that would allow us some time to relax and recover. On our first night, I ended up sitting right next to Madonna at a very posh restaurant near Portman Square. I was starstruck and couldn't sit still, having grown up with all her hits and dancing to her music.

But back to Mumbai. Publications like *Newsweek* and *Time* were doing cover stories on the emerging influence of India. All the food trend indicators pointed towards the relatively unexplored flavors and spices of the second-largest country in Asia and the fastest growing population outside of China.

I was sure that my newest concept would be the P.F. Chang's of Indian food and that by investing more than $1 million of my own capital and owning more than 50 percent of the company, I would protect my position to cash in once it expanded. It was a slam dunk. With not much deliberation, I kicked around a few names for my creation with Mark Miller on a sunny day while driving around Phoenix months before our trip to Mumbai.

"Bengal Coast," I blurted out.

"That's a great name," Mark said. "Stop there. You have it!"

With no financial pressures to stress me, I took my time doing research on the concept. I spent a lot of time and money exploring the best Indian cuisine in North America. There weren't many restaurants to use as analogs, but my two favorites were Vig's in Vancouver and Tabla and Tabla Bread Bar in Manhattan. Vikram Vig was in the same class, only on the other side of the continent and not as accessible. But I must have eaten at Tabla a dozen times with dozens of friends and family just to feed my enthusiasm for modern Indian cuisine that the stellar chef Floyd Cardoz was introducing. I became obsessed with Floyd's food and reached out to him through Mark Miller.

We struck up a friendship during my frequent NYC visits, and for a short period, we entertained the idea of moving him to Dallas to help with Bengal Coast. I smile as I look back on this now. Floyd was richly compensated and comfortable being a key player in Danny Meyer's Union Square Hospitality Group, Tabla's parent. It would have cost me dearly to relocate him and carry his salary during the time it took to open Bengal Coast.

But during our conversations, when it became apparent there was no way to merge my money with Floyd's skills, he did recommend a sharp young Indian chef who apprenticed with him. His name was Nevielle Panthaky, and he was a graduate of the Culinary Institute of America in Hyde Park, New York. He also suggested I investigate another chef in Boston: Anupam Joglekar. I was determined to meet them both.

First came Anupam. He worked at a fine Indian restaurant near The Commons called Mantra. I was immediately struck by his professional demeanor and assured sense of self. He went right to work creating all sorts of specialties from Mantra and other wonderful flavor combinations. My mind was made up before I left the restaurant. I offered him the job and was determined to move him to Dallas.

Confidence was a trait we shared, and though I never asked him what he was earning, I was committed to matching his salary and providing a seamless transition for him and his growing family of four at his earliest convenience. Anupam drank my Kool-Aid, and within a week, we had a deal. I was sure they would be comfortable in Dallas and its growing Indian population—another win-win.

Next, I set out to meet this hotshot chef who had worked with Floyd in New York. After some research, I found him at Estancia, a resort near La Jolla, California. I cajoled my old friend Mo Bergevin, my counterpart for Pei Wei in the Phoenix area, to join me in La Jolla by dangling the prospect of a few days of golf at the famed Torrey Pines Golf Course. The lure of great food and California coastal golf gained me a travel companion on my quest to meet this up-and-coming superstar chef.

Sometimes, you just know when you meet someone that you're going to have a long-lasting relationship. The moment Nevielle approached our table at the resort and flashed his award-winning smile, I knew I had my second hire. Deeply handsome and pristinely dressed in chef whites, he oozed confidence and unmistakable charisma. If that wasn't enough, the culinary creations he served Mo and me were beyond my wildest expectations. My mind was already calculating the riches we would earn together.

Pairing Nevielle with Anupam seemed like a no-brainer. I made him an offer on the spot, recruiting him at his place of employment. We came to an agreement, and within weeks, I sent him a check to move him and his wife Michelle to Dallas. They requested a "meet and greet" trip to the city to fortify their decision, but I knew that if I could get him there, he would come aboard. After all, I was a master recruiter, and I was determined to work my magic developing a concept I could show off with the best of Dallas!

I was on cloud nine, knowing that I would have two extremely accomplished and connected Indian chefs who would take the city by storm. It was as good a start at assembling a team as I could have imagined and would be the foundation that would catapult us all to fame and fortune. Anupam, Nevielle, and their families moved to

Dallas without a hitch, and we set out to learn about each other and get started in my large home kitchen to see what we could cook up.

That lasted a few months. All the while, I was paying both chefs their full salaries and paying for the ingredients required to prepare the dishes we were cooking daily. In retrospect, it was completely absurd, and I chalked it up to the dues I had to pay to create the next hit national brand. My entire house grew fragrant, thick with Indian spices and curries simmering on the stove daily for months.

I had secured a handful of investors to soak up the remaining 49 percent ownership block, and there were several tasting parties that flowed with various wines I was testing for our wine list. I was spending money as if it was in endless supply, which, of course, it wasn't. The testing came to an end, and the restaurant was being designed and built, operating from a budget that was generous by normal standards, but not outside that of the typical upscale venue.

In mid-2007, there was no Doppler radar warning us of the impending financial crisis that struck in the middle of 2008. So the checkbook remained open, and the add-ons kept piling up. In the meantime, I had hired a general manager and another manager, both of whom had worked for me at Pei Wei. With my team of four experienced, highly skilled managers, I was well equipped to open the next big thing in Dallas. At least that was the plan.

Never mind that I had already spent about $250,000 on travel, recruiting, moving expenses for both chefs, and salaries pre-opening. The dream was far from reality. No hammers had been swung; no walls had been erected. Another missed signal of the impending doom.

* * *

Continuous delays, poor project management, an inability to bring critical construction tasks to closure on the part of my contractor, and general fire drill-like activity led to a very disjointed and over-budget opening in January 2009. It was a less-than-ideal time to open a new restaurant, and all the buzz we had hoped to generate froze in the winter weather. Yet I was proud of the food, and despite conflicts with my contractor, the restaurant was stunningly beautiful and sexy.

But there simply is no way to get people to come out and try a new restaurant in Dallas in the dead of winter during an economic slump. The statistics are out there if you want to research them, but even P.F. Chang's experienced double-digit sales declines during this period. The toll it took on our industry was catastrophic. I did everything in my power to keep my dream alive.

But everything was stacked against me. Every time hope emerged, it was dashed by some unfortunate event. At one point, we thought we were handed a promising opportunity that we had hoped would turn things around. But even that was dashed by the weather. This was at a time when Indian culture was riding a high off the success of the feature film *Slumdog Millionaire*. It captured eight Academy Awards in 2009, including Best Picture, and the cast, along with their entourage, was making a stop in Dallas for a celebration that summer.

Because Dallas had a growing and very affluent Indian community, and because many of those affluent people had come to enjoy Bengal Coast, we were selected as the restaurant for the after-party. Our team was psyched. It would expose us to hundreds of local Dallas movers and shakers. Nevielle, Anupam, and our team spent countless hours planning a kick-ass after-party menu.

But a freakish summer storm blew through Dallas two days before the event was scheduled. The storm blew down the scaffolding and the stage that had been set up downtown for the *Slumdog Millionaire* extravaganza, washing out all hopes for the Dallas celebration. The entourage packed up and passed us by. It was just one of the many pins stuck in the Bengal Coast voodoo doll, but in the end, it proved to be the most lethal—though we somehow survived for a time, riding on the dedication of our loyal guests.

That is not to say we didn't have our successes, because there were many—part of the dichotomy of my predicament. We received a coveted "Best Indian Restaurant in Dallas" award by the *Dallas Observer* in their annual awards issue. They called Bengal Coast "inspired" and "startlingly fresh." The acclaimed *D*, our city magazine, declared, "The ambitious Bengal Coast shows much promise" and cited our décor as "dramatic and upscale."

And the bible of restaurant review sources, the *Dallas Morning News*, proclaimed, "Enjoy the attention of owner Mark Brezinski while it lasts. If Bengal Coast takes off like his other concept Pei Wei Asian Diner—and as polished as it is there's no reason it shouldn't—this could be the start of something big."

"Indian food had become a cliché, and it wasn't really mirroring the true Indian experience," says Chef Mark Miller. "Mark realized there were complexities within Indian sauces and palates that would be recognized by—and appeal to—the American consumer. There was an opportunity to use Indian flavors in a fresher way; a diversity, brightness, and intensity that he knew Americans would respond to. I supported him and believed in his direction and overall goal. It was innovative and smart."

The roller coaster ride was hard not to enjoy, but it was equally hard to avoid the nauseousness that came with it. And all of it was becoming impossible to sustain. We had no flexibility on the lease, regardless of how everyone was suffering during the recession. Our building was home to some very powerful and recognizable restaurateurs operating excellent restaurants. We did our best to band together to draw crowds. But the die was cast, and two of the restaurants—Silver Fox Steakhouse and The Club—closed before Bengal Coast.

I began to drink more each night at work and became sullener, ignoring most of my friendships. I became a lousy husband. I stayed too long at night, drowning in a volatile cocktail of attention and accolades while watching my bank account shrivel. I wandered, often aimlessly, from my restaurant to my car, full of self-pity. Not exactly the stuff of great marriages.

One and a half years into this venture, I realized it might break me. Though I had enough money to keep it open—and I stubbornly did so—I was forced to cut back on expenses. I closed for two days a week, reduced my management team to two, cut Chef Anupam loose, and designated Nevielle as both general manager and executive chef. We accepted any type of business we could get and even booked late-night disco-style parties in our atrium. It had nothing at all to do with modern Indian cuisine, but it paid the bills.

My ex-sister-in-law Teri Scroggins doubled as my bookkeeper and my day host. She worked magic with vendors and occasionally slipped me a one-hundred-dollar bill so that I wasn't walking around with no money. I had started Bengal Coast with somewhere between $1.5 to $2 million in the bank, and it wasn't long before that figure dipped below $100,000.

Jana and I had a stunning six thousand, two hundred-square-foot home in the upscale Preston Hollow neighborhood, and we had to cover a monthly nut of $20,000. We owned a half-million-dollar condo in Scottsdale as well. And as things continued to spiral out of control, my long-time accountant Lamont Grogan and minority investor and attorney Ira Tobolowsky begged me to cut my losses and close the business.

My answer? More drinking, more denial, and more poor decision-making. I started taking Ambien nightly just to get some sleep. I flirted with some of the great-looking guests that came into the restaurant, occasionally using the Bengal bathrooms for passionate interludes. Physical attractions become distractions. Guilty.

Something had to feel good regardless of how temporary it might be. Being an owner of a beautiful and occasionally busy restaurant, no matter how it might be failing, offered opportunities. I indulged.

I worked from just before lunch to dinner rush every night and avoided every responsibility a smart business owner would pay attention to. Teri would call me into our office to discuss ways we could manage to hang on. Inevitably, I would tell her that I would deposit more money from my personal account into the Bengal Coast account. Through her protests, I'd assure her I wouldn't keep doing that.

But it was the only way I knew to avoid the reality of my failure. I opened lines of credit based on my strong financial standing from my Pei Wei years. Yet I had no idea how I'd make those payments when they came due. Occasionally, to make payroll, I'd charge large Bengal Coast dining expenses on my American Express card, juggling payments to the restaurant against hopes for future sales. My card got canceled when they caught on to this scheme.

I managed this way for another year until Lamont called me one day and insisted I draft a plan to close the business. My sister Lois wrote me and reminded me that my business didn't define me and that this was just a part of my life journey. My team surrounded me with support, but the money had run dry, and I notified my landlord that I had decided to shut down my dream. At first, it was under mutually agreeable conditions—the recession was far from over, and the building was suffering from low occupancy. Fuck, everyone was suffering—office tenants, retail tenants, restaurateurs, staff, leasing agents. Everyone.

We agreed to give it one final week and announced a going-out-of-business celebration for those who remained loyal. My financial partners were unbelievably supportive. I had used so much of my own money that I never had to issue a capital call. Ever. We had a very emotional final night, got hammered, and faced the reality that there would be no more food cooked, drinks poured, reservations booked, or rushes to open the doors, hoping more guests would find us.

Closing a restaurant is a weird cocktail of grief, relief, deep emptiness, resignation, and panic. The ending was like falling off a cliff. I had to agree to walk away and leave everything in place, including plate ware, décor, and equipment. That would satisfy my lessee obligations even though I had seven years left on my lease. I did all that was requested, even agreeing to a security guard to walk through the space with me days later to ensure I wasn't removing anything.

I hung my head as I locked the door for the final time. It had finally hit me. I had nothing at all to do the rest of that day or for the foreseeable future. I wasn't broke, but it was time to scramble. I have never been—and probably never will be—as proud of a concept as I was with Bengal Coast.

The food, atmosphere, people, vibe, music, drinks, my team—all of it. It was as pure an original idea as I've ever created. I doubted I would ever top the stunningly beautiful Bengal Coast and its stellar team. I leaned against that door and cried. The failure was complete.

Of the lessons I learned over the course of my entire career, I absorbed the most in those two and a half years. These include being patient with site selection and being more aware of budget constraints and project timelines. The hardest part to accept is that I lost the money of those who invested with me. My head still drops, and that awful feeling comes back even as I write this that somehow, I let them all down. They trusted and believed in me—even if it was money they could afford to lose.

Bengal Coast was an innovative meteor across the Dallas skyline at an awful moment in history. It is part of my fabric now, an indelible imprint on my being. I was empty and as down as I've ever been. Yet the worst was still to come.

7

Aftermath. The Long Nightmare

My Bengal Coast adventure cost me $1.5 million. I was stunned. It never occurred to me that I could fail. It wasn't out of arrogance; it was just that the momentum that propelled me forward over the past ten years gave me the confidence that I would succeed. I had no Plan B. Bankruptcy loomed. But I had no idea how far off the rails things would get until one fateful evening.

"I've found the love of my life," Jana proclaimed, months before the final nail was in the Bengal Coast coffin.

"What the hell do you mean 'I've found the love of my life'?" I snapped. "What does that make me?"

"You scare the ever-loving shit out of me, Mark," she cried. "You're going to lose everything we have and all that we've worked for, and I just can't take it anymore. I want a simpler life, a small house, a white picket fence, and no money problems. You want more than that, and you'll risk everything to get it. I can't do that anymore."

I was stunned to hear those words from someone I had been with for fourteen years, married to for twelve, had traveled the world with, and who remained the love of my life. Maybe I should have seen it coming. Certainly, there were signs.

Once after closing the restaurant and arriving home late one night, I found Jana sitting on our bed—the light from her flip phone shining on her face—reading a text and laughing. She quickly snapped the phone closed. I didn't make much of it as she brushed it off as simply a funny text from a friend. Without skipping a beat, she told me it was from a man she had met on one of her trips to Wisconsin. A law enforcement officer of some type.

"Oh, you mean that guy you met at Tish and Scott's party?" I asked.

Tish and Scott were close personal and professional friends of ours. We all worked together at Brinker during Macaroni Grill's halcyon years, and the four of us eventually found ourselves working together at P.F. Chang's and Pei Wei. Jana worked closely with Scott and Tish in the development department.

But eventually, Scott and Tish decided they'd had enough of Dallas and moved to Scott's home state of Wisconsin. Because Jana still worked for them, she was taking quarterly trips to Wisconsin for departmental meetings. It was on one of those regular trips—for the annual department Christmas party that Scott and Tish threw at their home—that Jana met this guy who was a casual friend of theirs.

"Yes," she said.

He was also married, Jana shared. And I remember mentioning that I didn't think it was appropriate that he be texting her that late at night. But as was the case most nights, I simply had too much else on my mind to make much of it. The most important thing when your business is failing and you're hemorrhaging money is the next day. You never know what might happen. You live on hope, even when there's really no reason you should have it. It's the only nutrition you've got.

Another time, after stopping home between lunch and dinner shifts, I discovered a FedEx box on our porch. It was addressed to Jana. The return address indicated it was shipped from Wisconsin Dells. I brought it into the house.

"What's this?" I asked Jana.

She acted surprised but opened it while I watched. Inside was a Harley-Davidson baseball cap. I remember how quickly she lit up. A huge smile broke out across her face. I was puzzled. Then, she told me who it was from.

"He knows how much I like motorcycles," she said, that quixotic smile still firmly planted on her face.

Now, I was guilty of my own occasional flirtations in the restaurant business, some venturing well beyond sharing seemingly insignificant gifts, some way more personal. So, in that context, I made this out to be nothing more than some long-distance excitement for Jana at a time when I wasn't exactly putting much effort into exciting her. Over the course of the next few months, the fabric of our marriage unraveled. Jana eventually admitted that she and her friend from Wisconsin were more than friends and that conversations about her moving to Wisconsin were already underway. I was now dealing with a failing business, a listing financial ship, and a rapidly deteriorating marriage.

How could this be? I'd flown Jana around the world in first class, put her up in the best hotels, and introduced her to world-class cuisines. We owned a million-dollar-plus mansion, a condo in Scottsdale, and all that came with it. But it was that Harley Davidson baseball cap that seemed to make her happiest. How could I have been so blind to not see what really made her tick?

On Thanksgiving weekend in 2012, Jana packed up her belongings and prepared for the long trip to Wisconsin. She had no interest in negotiating or dragging our marriage through a long, drawn-out divorce. She knew our bones would be picked clean by lawyers and the courts. So, she quietly asked only for a few pieces of art we had collected and some dishes as she gathered her personal possessions.

Her new boyfriend was driving down to pick her up. He would rent a trailer, pack it with all her stuff, and that would be that. I made sure I was gone when he arrived. There was no verbal battle or sessions of blame. There was no hope of reconciliation.

She would file for divorce from Wisconsin, and she seemed eager to begin her new life with this law enforcement officer behind a white picket fence in Wisconsin Dells. She still lives there as I write this.

Her escape to Wisconsin, while sudden and disturbing, was the most courageous thing she ever did over the entire fourteen years we were together. I was always the one who made the decisions in our relationship, and now she was making the most important decision in our lives together on her own. In one sense, I gave her credit for finally being strong enough to do something bold.

In another sense, I'd never felt more desperate and alone—a desperation that led to me over-indulging in pain pills and self-pity, writing depressing poems, and watching hours of mindless TV. None of these diversions were familiar to me. I was in a whole new world created by some very bad decisions.

In any event, there was little time for commiseration. I was now jobless, close to penniless, and all alone. I'd finally reached the end of the rope, the end of my hope. I always knew crashing dreams would be agonizing. But I had no idea just how badly it would hurt and how deeply it would penetrate.

* * *

When you have a pattern, it's hard to accept an interruption in that pattern, especially if you're certain that pattern is permanent. For more than three years, Bengal Coast was the first thing I thought of when waking up and the last thing on my mind before drifting off to sleep. There were no vacations, no golf, no holiday celebrations with family, limited interactions with friends, and nothing that would qualify as fun or adventurous.

Once we vacated the Bengal Coast premises with the few things we were allowed to take, I handed over the keys, and heavy chains and padlocks were put on the doors. It was a not-so-subtle sense of finality. Jana and I had very little cash. I lost every penny of our savings. The issue was how to generate the short-term income we needed to survive until the "next big thing"—whatever that might be. It's not exactly typical of me to have a Plan B in place. I never thought I would fail, and I maintained the belief that I could sell the business right down to the point where I couldn't sell it because I was locked out.

We managed to do some income averaging accounting with my personal advisor, and we were able to claw back about $100,000 of tax payments to live on for a while. We sold our large house in Preston Hollow but broke even on the sale. We'd moved to a very modest house north of where we had lived, but it was a one-year lease. My intent was to get something going again and find more permanent housing.

But we were shaken by one bad thing after another. One of those things: I was served with a lawsuit by an attorney representing the landlord of Bengal Coast for the balance of the lease payments for the space. The balance came to $1.5 million, or $16,000 per month, for the ninety months remaining on the lease.

This wonderfully disconnected landlord, who lived in Malaysia most of the year, wasn't happy with me just leaving the restaurant and its equipment and furnishings in place. He was sure I had other money put aside somewhere, and he set out to find it. This cocksucker knew I had no more money, as I had revealed this to the property manager, but he didn't buy it. And when someone you owe money to through a personal guarantee decides to press the issue in court, the burden is on the debtor to prove there are no assets to satisfy the lease agreement in whole or in part.

When the person filing the lawsuit believes you have more money, they can force the debtor's hand through bankruptcy. The goal is to scour for saleable assets to pay off the obligation. The only thing protected is your homestead and the necessities you need to live. The process is humiliating. I had nothing hidden, no money put away for a rainy day, and owned not a single thing that wasn't clothing or furniture or laundry soap. It took months, but ultimately the bankruptcy court determined I was incapable of paying off my debts.

To make matters worse, our lease was coming to an end. We had to move. Jana agreed to help with the process, but it was obvious that this was the last thing we would ever agree on. We completed the move not long after Jana announced she was leaving me that Thanksgiving weekend. To add to this sad drama, I now had to find the money to pay for my bankruptcy attorney.

It was through the generosity of a few very good friends that I was able to manage to not only pay my attorney but also to pay my bills—my car payment, insurance, utilities, cell phone, and so on. On one desperate occasion, I put in a call to a close friend. I was driving to my bank to withdraw my requisite allocation of twenty dollars, and I broke down. This friend had always been smart with his money, and while he had offered to help, my pride kept me from accepting. But I was at my wit's end. I called him and erupted in tears. I was older than he was. I had a more public standing in the industry. But I wasn't as smart as he was in the financial realm.

"Can you help me, Tony?" I managed to somehow say. Between tears and pauses, I explained my dilemma.

Had he not offered—unconditionally—to help, I truly do not know where I would have turned. In bankruptcy, you lose all your credit cards and the ability to get a loan. Friends and family are the only options, and I'd already ruled out family.

It's a gigantic fall from where I had been just a few years earlier with seven figures in the bank. Now, I was managing every single expense down to the penny. It was eggs for breakfast and dinner, no eating out, and fast food instead of steaks. I had to strategize to keep gas in my car, aligning the miles I had to travel with twenty-dollar ATM withdrawals timed several days apart. If someone invited me to lunch, I mostly declined because I didn't want to presume they would pay, and I knew I couldn't.

I withdrew from even my closest friends and, for weeks, didn't leave the house except for grocery runs and to walk my two Labradors. I experienced severe pain in my hips. They had been slowly deteriorating over the years, and I was diagnosed with severe arthritis on both sides, making hip replacement inevitable. I had a prescription for Ambien continually renewed since my Bengal Coast days and did my best to convince my doctor that I wouldn't get any sleep without it.

Now, with the hip pain worsening, I was adding painkillers to my sleep cocktail just to get a decent night's sleep. The pain pills stopped being just nighttime relief, and I was soon taking more than the prescribed dosage. For months, my days consisted of waking up, eating breakfast, walking the dogs, watching TV, and doing whatever I could to search for work.

You must remember this was on the backside of the financial collapse and the worst economic recession in modern times. Very little was being done in the restaurant business that was new. The only thing keeping me from adding homeless to my resume was the support from friends.

Things got so bad that one morning, I decided to reach out to my professional network and essentially beg for work. The email list included about thirty people, and when I hit send on that email back in 2013, I drew a deep breath. Two friends reached out to assist:

Randy Dewitt, and my former employer, Vincent Mandola, who would pass away suddenly in the spring of 2020.

* * *

Randy Dewitt was a longtime colleague who had himself come close to financial ruin until he one day decided to convert one of his Rockfish Seafood Grill locations into a new concept in the "breastaurant" segment. He surveyed the dining landscape and discovered there was one concept that seemed to have no competition—Hooters. He set out to create that competition with a mountain lodge-themed venue he smartly called Twin Peaks.

That turned out to be a huge national success and it gave him the juice to branch out and experiment with other concepts. One of those was a taco joint he called Taco Libre, which eventually morphed into Velvet Taco, which I discuss later in this book. He had watched closely when I had developed Tin Star and decided that maybe I had another taco bullet in my gun. That earned me a modest $5,000 monthly fee. I started this new consulting gig after about six months of no work.

Mandola wanted help with his fledgling Italian fast-casual concept, Pronto Cucinino. I had helped him and his family launch Pronto many years earlier when I was still with Pei Wei. He finally saw the possibilities with Pronto but hit a ceiling on development. The problem was that the work he had for me was back in Houston, and I was living in Dallas. We worked out an arrangement for me to travel twice a month to Houston for one week of work at a time, two weeks a month, for another $5,000 monthly fee. He rented me a garage apartment. I made the bi-monthly trek for about a year.

Once this work dried up, as consulting work always does, I needed to go on the hunt again to keep from drowning financially. Even good friends and family don't continue to offer endless relief.

I did some more networking and happened to connect with Micky Pant, global CEO of Yum! Brands Pizza Hut International and a big fan of Bengal Coast. He introduced me to one of Pizza Hut's largest franchise groups based just outside of London in Borehamwood, England, Pizza Hut UK Limited, or PHUK. That group was seeking assistance with changes to their Pizza Hut full-service restaurants in the UK.

I was able to secure that consulting gig for $7,500 a month. My work wasn't to re-program the Pizza Hut process but to give the PHUK team input on issues like design and menu features permitted under the franchise agreement. When business is on the decline and being cannibalized by competitors and changing trends, the money needed for wholesale changes simply isn't there. They were hemorrhaging cash and looking for inexpensive fixes. That's what I was charged with providing. My proposal was approved, and work began.

One problem: I had no credit card for travel and no money for incidentals—not even the train fare from London Heathrow Airport to central London. But not going forward was not an option. So, I swallowed my pride and called my ex-wife Jana. When we were married, we obtained a separate credit card for her, and since we were so well financed, she received a Visa credit card with a $25,000 limit.

I asked her if she would add my name to that card and allow me to use it for my consulting work. I promised to pay her the monthly expenses I racked up along with a 10 percent fee for her trouble once I was reimbursed. She agreed. This saved my ass because, without that card, travel was out of the question. But this was tricky. I had to present a professional image while not revealing how desperate I was.

So I had to come up with a way to convince PHUK that it would be easier on them if they simply purchased my airline ticket and offered a credit card authorization for my lodging. This way, I reasoned, I wouldn't have to submit any expense reports for my travel and lodging. They agreed to arrange all my travel and my hotel accommodations.

On my first trip, I was unable to get the credit card in time from Jana. I was traveling without that parachute and chanced being exposed for the penniless professional I was. I arrived in London on a ticket purchased by PHUK and had enough funds to take a train from the airport to Borehamwood. But that was about all I could stomach financially. My debit card had less than a $500 balance—hardly enough for a week in England.

I presumed PHUK had made all the arrangements for my one-week stay at the hotel. But when the desk clerk looked up my reservation, there was no notice of third-party billing and no credit card on file. They had made the reservation but forgotten to authorize the stay on their corporate account.

I hid my panic from the desk clerk well. I casually convinced him there must have been a mistake and suggested he hold my debit card without charging it as I cleared up the matter. I tried to remain calm and engaged the desk clerk in small talk as I waited anxiously for the return email from PHUK. It wasn't to come. I had just enough money in my account for one night and suggested to the clerk to let me at least check into my room as we waited for a response from PHUK.

I left my debit card with the clerk and said I was going to take a shower. It was the most panicked hour of my professional life. The walls in my small hotel room were closing in on me. I was sweating profusely and shaking uncontrollably. The persona I needed to project and the vision in my mirror could not have been more juxtaposed. I'm not sure, but looking back, this may have been when I reach that proverbial bottom of the barrel.

I needed a miracle to get out of this jam. When the phone rang in my room, it startled me. It was the desk clerk. He informed me that he had just gotten a call from PHUK and that my stay for the week was taken care of. I immediately took out my laptop and composed a thank you email to my employer. I'd narrowly escaped disaster that time, but I get chills just writing about how it felt sitting alone in that nondescript hotel room in a place I'd never been before.

Another issue I had with this adventure was air travel. I am six foot four and about 275 pounds, and the flight from Dallas to London—about ten hours—is less than comfortable in coach. But the cost of a business flight was something PHUK wouldn't pay for. I somehow convinced them to purchase two coach seats side-by-side to accommodate my size. I settled in with my work spread out on one seat and me in the other.

It didn't take long before my flying status improved. I was soon able to get upgraded on most flights, and I stopped asking for two seats. I dug in and believed that my work with PHUK over two years was some of the best I'd ever done. Once I gained momentum with my work and cemented my income, I began to see the light at the end of this terrifying tunnel.

* * *

Nothing came easy during this dark period. Hanging heavy over my head was that I still owed friends who had helped me with substantial sums of money. And though I was now able to pay my monthly bills and Jana her 10 percent, I was far from out of the woods. I wasn't paying quarterly IRS payments; I just didn't have enough cash for those. I knew I couldn't avoid this forever.

On top of that, Jana called me one day and said she was getting married and that her fiancé didn't want her to continue carrying me on her credit card account. She was changing her name and felt this was a good time to end our arrangement. Though I was making enough money to pay my bills, I was still a pariah to credit card companies because of my bankruptcy. I had to find a quick solution.

By now, you must be getting the sense that I do my level best to solve problems myself and avoid being exposed for what little I had to my name. Though my small circle of friends might describe me as a stubborn survivor, the truth is I simply learned how to swallow my pride and try my best to keep a lid on the panic I felt daily. I dialed my oldest brother Dick—who I'm sure had no idea of my predicament—and offered him the same credit card deal I had with Jana.

Dick was retired and living very comfortably. But the prospect of a few hundred more dollars a month was appealing to him. He took my offer. I never shared with him just how desperate I was. Among my siblings and I, none of us ever really had a close relationship, and we've always lived long distances from each other. My siblings knew me as a wild free-spender with no children, and I did nothing to disabuse them of that notion. I simply thanked him and anxiously awaited my new credit card lifeline.

When it arrived, I didn't skip a beat. All my reimbursable business expenses were now netting Dick extra cash that he never questioned but surely enjoyed. You never live down going through bankruptcy. It might go away, and you might get credit cards again, travel, and live comfortably. But you never look at your friends or family in the same way. The stain of exposure, the shame, and the darkness inevitably becomes a part of your fabric. You might laugh often and find diversions in dating and other forms of recovery. But you never forget.

The process of crawling back is slow, and the ability to share even slower. I lost contact with many of my friends over those years and retreated from others who encouraged me to get back into the mix. Instead, I completely immersed myself in work and spent more time writing, often poetry and long emails to the few friends I continued to correspond with.

I'd say this period lasted the better part of three years, and while few of my friends were aware of this, I wished I'd done better at letting them into my life. I knew some of them simply didn't know how to reach out in a way I could handle. I was always the guy who picked up the dinner check, had plenty of money if anyone needed help, traveled first-class, and wore expensive clothes every day to set an example for his team. You could say that I lived beyond my means. But I'd tell you that netting $60,000 every month allows you to define your own means. I never shied away from spending, and I never turned down a friend in need. And I never imagined the possibility that I'd someday be broke.

I became the guy who was eating hot dogs at home for dinner, was taking no more than twenty dollars at a time out of the ATM, and had spent not a dime on clothes in years. I will never be either of those guys again. My peace with my mistakes and my acceptance of who I am is complete. Buddhists say, "Wherever you go, there you are." In some ways, I am more comfortable now with where I am than ever before, having survived the dark aftermath years. Or so I thought....

8

Zinsky's. A Good Catholic Boy Opens a Jewish Delicatessen

Growing up in New Jersey in an ethnic neighborhood with access to just about any cuisine you could imagine, I gravitated towards deli fare. I was fascinated by the huge sandwiches delis were known for. Our easy access to New York City and the intense deli culture there—Stage Deli, Carnegie Deli, Katz's Deli, and Zabar's Market on the Upper West Side—only fed my appetite. Of all the foods I've ingested in my sixty-eight-plus years, the Reuben sandwich (with pastrami!) may hold top honors.

My experimentation started early, when I was a boy, during our requisite family visits to Wildt's Delicatessen on the way home from our weekly Sunday mass service at Holy Angels in Singac, New Jersey. Fresh onion rolls, Kaiser rolls, Italian loaves, and pounds of thinly sliced deli meats comprised our weekly post-church lunches at home—on those weekends, we didn't buy sacks of White Castle hamburgers! I can still smell the onion roll aromas wafting from the bag, anticipating a sandwich made from the shaved spiced ham,

Swiss cheese, pickles, and a heavy dose of Hellmann's mayo. It's a habit that lives with me to this day.

As I grew older, it was always my goal to eat at every Jewish delicatessen I could find in any city I visited. I achieved that goal. In Houston? You must go to the legendary Kenny and Ziggy's. In the Detroit area? You better find Zingerman's in Ann Arbor. Milwaukee? Head to Jake's. Find yourself overnight in Indianapolis? Shapiro's is a must. Nothing better in San Francisco than Deli Board. Find it. You won't be disappointed! The #19 at Langer's in Los Angeles? King of the hill!

So it was no surprise when I was eventually smitten with the deli bug. I found—or maybe it found me—an opportunity to create a neighborhood deli in the heart of one of Dallas's most exclusive neighborhoods: Preston Hollow. I had friendly relationships with two Dallas deli moguls: the Gilbert family, who operated Gilbert's, and Larry Goldstein, owner of Bagelstein's.

As luck would have it—and as Bengal Coast was dissipating into oblivion—I was approached by Goldstein for some advice. He was having problems with a Bagelstein's location in Preston Hollow. The new, smaller location wasn't working.

Goldstein shared that he was thinking of closing the location and wanted to know if I had any ideas on how I could take over the business and get him out of his lease. It was an expensive space, not overly big, but it was already set up as a deli. My wheels started turning. But I was in no position to launch a new venture like this without financial backing.

The first person I turned to was fellow northeasterner and successful Dallas restaurateur Jim Baron, owner of Blue Mesa Grill. Baron was a well-connected guy in both the Jewish community and the Dallas restaurant industry. We had never worked together but knew each other from Dallas's restaurant scene.

After a few meetings, we figured we could make it work. We formalized our plans and came up with names and menu ideas. Our first objective was to come up with a moniker that merged both of our names. Using the first letters of both of our last names, we came up with B&B Deli. But the people we ran it by said it sounded too much like a bed-and-breakfast.

Then I came up with the idea of using the last part of my name, "zinski," as the foundation for the name, changing the "i" to a "y"—more typical of the spelling among Jewish immigrants from Russia and Eastern Europe. I proposed the name "Zinsky's," and while Baron and his wife Liz first balked at elements of their name being left out, the reception to this from investors was far better than anything we could come up with. Zinsky's Deli it would be.

We raised $750,000 for a limited partnership backed by ten investors—all but two of Jewish heritage. We hired a great local chef, Lex Berlin, who I had worked with during my Macaroni Grill days. Josh Garcia was hired on as general manager.

Our team embarked on a whirlwind tour of all the New York delis we could stomach. We did our best to ingest as much chopped chicken liver, matzo ball soup, lox and bagels, and Reuben sandwiches as we could cram into a three-day trip.

Our goal was to experience the food, culture, and requisite elements of the typical New York City delicatessen so that we could incorporate these elements without trying to copy any single spot. The worst thing you can do—and it's a common mistake among restaurant developers—is try too hard to copy something. The result is often something that is neither original nor as good as the place you are trying to imitate. It comes off as a hodgepodge. People notice.

* * *

My vision for Zinsky's was to evoke an authentic sense of New York deli culture in a setting that was comfortable, with the expected menu options like matzo ball soup, corned beef, pastrami, bagels, and free deli pickles on the table. But we also featured an expansive menu covering breakfast, brunch, lunch, and "blue plate specials" for early evening diners.

Fast-forward.

When we opened, the reception was enthusiastically positive. Being in the heart of one of the city's most expensive neighborhoods at a very busy intersection didn't hurt—location, location, location. There was pent-up demand, and we were full for breakfast and lunch every day, with Saturday and Sunday being especially crazy. I felt compelled to escape the chaos of busy weekend shifts. The noise was deafening, and the activity distracting. It conflicted with my need to step back and analyze things.

To stand at the expo counter on a busy lunch shift and watch food piling up with the madness of the plates hitting the deck and servers calling for food was just too much for me. Exacerbating all of this was the fact that eggs get cold more quickly than other dishes— like bowls of noodles, thick steaks, or dense soups. I just didn't have it in me anymore to know that people treat eggs like gold and are not shy about complaining. I would have argued myself out of business if I had to listen to one more person describe what they thought was "eggs over medium" and ask for a re-cook.

My days of reveling in high-energy restaurant stimulation had passed, and I was more than willing to hand that off to others. As for Baron, he was virtually nonexistent. We had agreed that he wasn't really needed to help oversee a business of this size. His role, as we'd clearly agreed to at the beginning of our partnership, was to oversee the business end of the operation: paying bills, managing the books, and doing payroll—all offsite. We didn't need or want any other input. And neither Baron nor his wife Liz did much for the business other than paperwork and managing social media and PR strategies. It was a formula that worked. Or so it seemed.

Once the business leveled off, we held more meetings with investors and hosted strategizing sessions to address emerging challenges. We weren't drawing a robust dinner business, and our weekday breakfast sales were slackening. We decided to ratchet back the evening hours and do some more PR to promote breakfast.

Baron wasn't an active participant in these strategy sessions, yet we sensed he wanted to be more involved since we saw him in the deli more often than when we first opened. His visits to Zinsky's grew more frequent, and he offered more input on the menu and operations. This was in stark contrast to his hands-off approach during the opening process. It felt odd. My antennae went up. My suspicion of Baron increased.

But he was a partner, and we took his sudden interest in the day-to-day operations mostly in stride. He insisted we bring someone else on board in a leadership role: a guy who wasn't particularly friendly or experienced. He also knew nothing about the deli business. I labeled him "Jim's spy." Little did I know how right I was.

This guy was not tuned into hospitality or delis, and he seemed like a gefilte fish out of water. He was a chain-smoker who spent more time in the parking lot than in the restaurant. He showed no interest in being part of any team or in understanding the business. An oddball. Something was up.

One morning, Baron asked me to meet him at the local Corner Bakery. I thought it strange that he wanted to meet at what was essentially a Zinsky's competitor. I figured I was in for a surprise. I wasn't disappointed.

Zinsky's had been open for about six months, and though sales were good, and our reviews had been strong, we all felt we could do better. Nothing out of the ordinary for a new restaurant, and, by my reckoning, we were well ahead of schedule. I had a 10 percent stake in the partnership and was hoping my share might eventually provide some income. But based on what transpired at that meeting, it was clear that would never happen. Baron asked me to step aside and let him and his "spy" operate the business and oversee the management team.

It wasn't a complete surprise, and what I'd learned about Baron during the opening wasn't entirely positive. He was very shifty and not entirely honest. Besides the unwelcome presence of his spy, we found it nearly impossible to get any feedback from his administrative staff, and he never spent any time with the restaurant staff to build culture or morale.

My team of managers and I were used to weekly reports, vendor payment summaries, and direct communication with the support side of our businesses. But we never received any financial statements, and there was a feeling that Baron wasn't interested in what any of us thought.

One of our local vendors, who provided us with baked goods, started complaining that they weren't being paid. Another vendor, Mark Sheckter, who supplied us with freshly baked rugelach (a traditional Jewish pastry), even came into the restaurant one day screaming at our manager. He claimed we were all a bunch of crooks, and he wanted our guests to know how dishonest we were. He said we continually promised to make good on invoices and then reneged.

"These guys don't pay their bills," he shouted. "You can't trust them!"

None of us knew any of this. More vendors started coming forward with complaints, and it became clear to me that the financial side of Zinsky's was operating in a questionable manner. I knew that restaurants sometimes had to delay payments during slow periods. But Zinsky's was doing well. We should have been current on all our payments.

Baron never gave me access to the books. Was this clusterfuckery due to mismanagement? Incompetence? Or was Baron skimming from Zinsky's to subsidize his other restaurants in a kind of restaurant Ponzi scheme? I don't know. But to me, our meeting at Corner Bakery meant that Baron was afraid I was closing in on the truth. That's when I decided to plot my exit from something that might come back to bite me hard later.

* * *

After speaking to my attorney Ira Tobolowsky, I planned to turn my 10 percent interest in Zinsky's over to Baron in exchange for a signed letter of indemnity. That letter would shield me against any liability or obligations after my departure. Since I hadn't invested any capital—only sweat and expertise—I would lose no money and had no financial risk. Baron willingly signed the letter to get me out.

But Ira made one small—but very significant—oversight. He dated the letter from the day of our negotiations, not the date of Zinsky's formation. No big deal, right? I was out and had in my hands a signed letter stating no one could come after me once I'd escaped any potential Zinsky's mess.

About four months later, while I was working at home, the doorbell rang. Doorbells can mean many things. They might mean good news, neighbors stopping by with fresh baked goods to share, a mailman bringing packages that needed signatures, or maybe just someone returning your lost dog.

Not this time. I opened my door and found a stern-looking gentleman in a suit. He immediately handed me a business card. On that card was the official seal of the United States Department of the Treasury.

I invited him in, but he said he'd prefer we talk outside. He immediately asked if I was the Mark H. Brezinski that was one of the general partners of Zinsky's Delicatessen, LLC. Yes, I was once the general partner, I explained, but I had exited that company about half a year ago. The agent waited outside, while I hurriedly rushed into my office to produce a copy of the letter of indemnity. He then informed me that Zinsky's was under investigation by the Treasury Department for nonpayment of payroll taxes to the tune of $100,000 before interest and penalties. Because records showed that Baron and I were the two general partners, we were each liable for half of whatever amount the government deemed we owed. The agent at my front door suggested I get an attorney. Just what I needed. More legal fees.

As I closed the door, I was instantly overcome with panic and dread. What had Baron gotten me into? I sat down at my desk and immediately composed a lengthy email to him explaining what had just happened and what I expected him to do: Step up to the table and clear my name. Steam was rising from my head as I banged out that email. Fucking Jim Baron lived up to all my suspicions.

I also contacted Ira. He did his best to calm me down. He assured me that my obligations would be minimal given the letter of indemnity. Baron casually and dismissively responded to my email by claiming that, on the advice of his attorney, he was not required or allowed to speak with me.

I'm not sure what I was expecting from him. But I was certain he would somehow get stuck with all tax obligations resulting from his nefarious bookkeeping and that I would skate without further financial damage. Justice would prevail, right?

Wrong.

As a partner, I was ultimately held accountable for my share of the taxes and penalties *before* the date of the letter of indemnity—an amount that came to $40,000. With legal fees, that sum rocketed to $50,000. This drama came at the absolute worst time. I was amid bankruptcy proceedings from the Bengal Coast disaster, struggling to find more consulting income, and dealing with excruciating hip pain. Now, I had the Treasury Department—not exactly a flexible bunch—breathing down my neck for something for which I had no direct responsibility.

I have not seen Baron since that meeting at Corner Bakery nor heard from him since his last email. But I was so infuriated with Baron that I conceived a scheme to buy a barely running junk car, drive it to his house in the dead of night, and rev it up before plowing it straight into their front door. I would then remove the key and quietly walk away, leaving them to figure out what the fuck happened.

That was my plan, and I was more than ready to carry it out. But I came to my senses and let the fury pass. The investors I gathered for Zinsky's—moneyed people who trusted me—never received a dime of dividend disbursements. It's hard to imagine my name wasn't sullied.

Zinsky's closed a year after I was removed. To this day, whenever I hear the name "Baron," the sound spreads over me like a bad rash. I'm overcome with the urge to compulsively scratch until I chafe, or any hint of the name dissipates—whichever comes first.

Jim Baron may be the most deceitful and dishonest individual I have ever worked with in my forty-plus years in the restaurant industry. How he can wake up every day and look himself in the mirror without a pang of remorse or guilt is beyond me. I was advised against overusing the "F" word in this chapter because it might lose its impact. But I won this battle. Fuck you wherever you are, Jim Baron. I hope our paths never cross again—for your sake and mine.

With Chef Mark Miller outside a temple in Bangkok, 2008

Mark with Christophe Poirier and Yasmin and Braden Wages
at the opening of Bahn Shop, Dallas, 2015

Closing night, Bengal Coast, Dallas, 2011

Cover of Bengal Coast brochure, 2009

Grand prize, Fork Fight, Trinity Groves, 2013

Cheeseburger Katsu Sando

Eating out was my dad's passion, too. This is on the
porch of the Smithville Inn, Smithville, NJ, 1980

Mark's junior year at Kinnelon High School (#12), Kinnelon, NJ, 1970

Mark's first restaurant job flipping burgers at Anthony
Wayne Charcoal Grill, Route 23, Totowa, NJ, 1971

Mark with Chef Morgan Hull on the cover of *Restaurant Business
Magazine* as a "Concept of Tomorrow" in 1999

Mark with Pete Botonis and Shannan Metcalf-Perciballi, the three first hires of Pei Wei Asian Diner in 2000

Mark with his ex-wife, Jana, dining in STK NYC in 2006

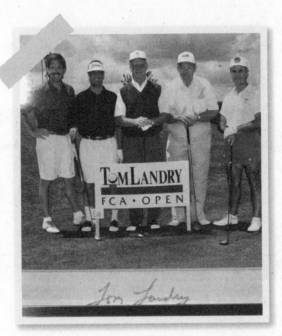

Mark with his Macaroni Grill mentor Rick Federico—to Mark's right in
the photo—at a Tom Landry Dallas Golf Tournament in 1992

Mark's bond with his mom started early as shown
here, holding hands at two years old

At the opening of Pei Wei, Dallas, 2002

As President of Marugame Udon, 2020

9

Velvet Taco. Lightning Strikes Twice

One of the aftereffects of getting back into more creative and productive restaurant work, which was now happening with my uplift during my stint with Pizza Hut UK, was the boost in confidence I began to feel. I'd never lacked confidence, though, for sure, the Bengal Coast failure lingered heavily in my psyche.

Inside, I began to feel worthy again—worthy of being in the company of my peers and my closest friends, pursuing the latest trends. It's hard to explain, but part of me has always felt as though I were an artist. Some things I see clearly as an original vision on my canvas as I develop a concept alone or in collaboration. Such was the case when I was asked by Randy Dewitt and Jack Gibbons of Front Burner Group Dining to join them on a project they initially called "Taco Libre."

Randy had answered my broadcast email I mentioned in the Aftermath chapter, and I was fortunate to be able to work on this new project concurrently with my Pizza Hut UK work. After we agreed on a consulting fee—which eventually turned into a partnership and equity position—Randy started sending me his palate of ideas and included me on the documents and emails shared with the team.

That team was a veritable powerhouse of the industry: Randy, Jack, Scott Gordon (their de facto CFO), Chef John Franke, the design team of Royce Ring and Alex Urrunaga of Plan B Group, a firm that would also lead the Taco Libre "DNA" development. When we got together in a room, the accumulated experience represented exceptional depth, including stints at Brinker International, TGI Fridays, Pappas Restaurant Group, and others. We were a dream team that became one of the most successful fast-casual stories over the previous decades of restaurant development.

The inspiration for this yet-to-be-created Taco Libre project was the "set my taco free" motif, essentially directing our group to think outside the usual taco box. Traditional street tacos were all the rage in Dallas at the time, inspired by the simple tacos found in many small cities dotting the Mexican landscape.

When I say traditional tacos, I mean soft tacos on homemade white or yellow corn tortillas, inexpensive proteins, and simple toppings like chopped onion, cilantro, lime, and salsa. The Mexican street-food influence in Texas was obvious, ubiquitous, broadly accepted, and appreciated.

There was a parallel taco trend started by chains like Torchy's Tacos out of the uber-cool restaurant scene in Austin, Texas. It was grabbing local attention as they entered the Dallas market. Torchy's wasn't a traditional taqueria. Instead, it focused on a wider palate of flavors, with creations like the "Dirty Sanchez," composed of ingredients that had ties to tradition but weren't bound by them. The creators intentionally wandered off the reservation a bit but without leaving the state.

Essentially, there was a taco street fight building, and Randy and his team wanted to get in there and mix it up. Bolstered by their incredible success with their Twin Peaks "breastaurant" concept and the return on investment that comes with such success, Randy and the team didn't skimp on development capital. Assembling this dream team was a strong indication of their commitment.

During my research for Taco Libre, I found myself unimpressed with Torchy's lack of culinary focus. Some of their taco recipes simply weren't very good and represented a concerted effort to be cute or different. That was a path I wanted to avoid. Instead, I clearly saw an opportunity to be innovative without being earnestly irreverent. The whole Torchy's shtick seemed forced . Quality always wins over cuteness; my focus was on quality.

We were essentially running in place until we decided that maybe the name Taco Libre was handcuffing us. Plus, there was a trademark conflict with an already existing brand. However it happened, our group turned to Chef Franke and me to begin developing tacos to sample while we brainstormed names. I've always felt very comfortable in kitchens during this kind of development, and what was difficult to describe in words to our team was easier to demonstrate with tastings. Though not a chef, I consider myself a trained consumer on top of culinary trends, something I take great pride in.

Really great food at a value price seems like a simple formula. But far too many people I have worked with have managed to fuck up that simple equation, letting their egos get in the way. Chef Franke was not one of those people. He and I worked well together, and we started to meet in the kitchen at one of Randy and Jack's restaurants: The Ranch at Las Colinas, just outside the Dallas city line.

It had a huge kitchen and plenty of space, though it was extremely busy, and we often had to fight for space to work. John and I started by sharing ideas, and soon, the boundaries were widened to include any ingredients we thought would work—not forced or manufactured, just great flavors that melded into something we personally would eat and would be proud to serve. We strived to have all our tacos predetermined by build instead of the more common build-your-own taco approach.

We also wanted a broad mix of proteins, from beef and pork to chicken and seafood plus vegetarian options and all-day breakfast tacos. Our quest was to keep our menu executable, with a slate of twenty variations that incorporated unique flavors that weren't yet in the Dallas market. Sure, there was a nod to some traditional offerings but with our own twists. The skirt steak taco would have white queso, grilled red onions, and portobello mushrooms, with a garnish of micro-oregano, for instance—nothing funny or trendy about it.

That became our mantra: no-nonsense tacos with globally inspired combinations that were unique but accessible. That's where my background in Asian food came in handy. We began having tastings and could see the group starting to get it. Our vegetarian option employed a roasted tomato chutney with grilled paneer (Indian cheese) and basil. The buffalo chicken taco—a nod to my Tin Star days—had an agreeable spark of heat, blue cheese sauce, and celery leaves.

When we were done and had our base menu of twenty tacos, it was clear there wasn't anyone in Dallas doing anything close to what we'd come up with. We were both regional American—a cheeseburger taco—and global—a fresh ahi tuna taco with pan Asian influences served on a fresh lettuce leaf. We had paid our respects to the Mexican tacos with a roasted pork taco with arbol chile ancho and homemade salsa and a breakfast taco with queso, sausage, and roasted potato. Our recipes were uniquely distinct, so we weren't easily confused with any number of local taco joints.

One day, during our weekly meeting, Randy said he'd come up with a name he wanted to run by us. He'd been talking to friends at a party about the new concept, and someone mentioned that the tacos sounded luxurious, like the fabric velvet. A light bulb went off in Randy's head.

"Why don't we call it Velvet Taco?" he asked.

The assembled heads shook and nodded, and after some spirited debate, the name stuck. But there was one problem. If you look up Velvet Taco in the Urban Dictionary, you'll discover that velvet taco is slang for vagina. To this day, I cannot tell you how many of us in that room were aware of this, but we knew soon enough after we shared the name with others.

It was a risky call, but the name wasn't protected or in use by anyone else, and the Plan B team set out to logo-ize it, steering the name away from any affiliation with the Urban Dictionary term. We managed to become a spot for luxurious tacos, and any concerns with blue terminology dissipated.

* * *

I began this chapter by stating that Velvet Taco helped me feel worthy again and that I had started to believe that I belonged in the company of other successful executives. Velvet Taco was instrumental in that journey, and soon, I was feeling the self-assurance I had during the halcyon days of Pei Wei, Macaroni Grill, and Canyon Café.

As we got deeper into development, it was clear that the team was listening to me, and though nothing formal was ever stated, I felt I was the leader of the group. I'd managed to hire an excellent general manager who'd worked with me at both Pei Wei and Bengal Coast as well as Banh Shop, which would follow Velvet Taco—a young, incredibly hardworking guy named Josh Garcia.

Josh became instrumental in organizing work at Velvet Taco, and his experience with exhibition kitchens and fast-casual restaurants made him a favorite of our team. His presence also helped cement my leadership role because, in the restaurant development landscape, an accomplished general manager who knew his way around the kitchen made the work of Chef Franke and myself even more effective.

The three of us took over development work and were responsible for kitchen design, the final menu, and the operational processes. To have a great idea is a good start, but there are lots of great ideas out there. A great idea led by a great team is the sweet spot. Josh Garcia was our best insurance policy, and everyone—from Randy to Jack to the Plan B guys—knew it. He was like the jockey that we hired to ride this thoroughbred into the winner circle at Churchill Downs on Derby Day!

Randy was able to secure us an excellent—albeit smallish—location in an area just north of downtown called Knox-Henderson, a nod to an intersection where the names of the streets changed as they crossed over Central Expressway, a major Dallas artery. The spot had been a Church's Fried Chicken restaurant for nearly twenty years and sat on a corner with minimal parking next door to the original Dickey's BBQ. Roland Dickey, the owner of Dickey's, owned the land under that Church's location and was a fan of a street taqueria taking over the plot.

It was the steal of the century real-estate-wise as the exposure on this street corner was immense. It was the gateway to one of the area's hottest and busiest nightclub and restaurant scenes. It was teeming with people all day and night, so the opportunity for consumer capture was as good as any location in Dallas. That's the good news.

The bad news was that the building was a wreck—raze-worthy—and needed a costly overhaul. The Plan B guys were as good as anyone in the design business, so the challenge fell on them to make lemonade out of lemons. My role was to make sure the kitchen design aligned with menu demands and that tiny one thousand, five hundred-square-foot building. Our little taco concept was taking shape.

One of the most difficult aspects of opening a new concept with no history or organization is that everything, from the employee manual to the recipe book to the training materials to the food order guide had to be developed from scratch. Fortunately, John and Josh both had deep experience with this work, and they attacked it with vigor. As construction came to a slow and painful close, the excitement and anticipation were enormous.

The attention we were getting was overwhelming. As opening day approached, we all felt very ready for what was about to come. Or so we thought....

* * *

All hell broke loose shortly after our start, and our tiny parking lot turned into a scene from Times Square. People from all over were discovering Velvet Taco, and our thirty-seat dining room was busting at the seams. It was unlike anything I'd ever experienced. Our entire team was shocked by the demand and acceptance of what we'd produced. It was as if we'd created a Frankenstein's monster, but he could talk eloquently, cook with aplomb, move effortlessly, and had the hospitality sense of the best hotel concierge.

There were long lines, day and night, with the kitchen humming pretty much twelve hours straight. People filled the dining room, waiting for seats while hovering over others, scarfing down $3.50 to $5.00 tacos and quaffing down frozen margaritas. Unlike most such concepts, we offered only one ultra-premium frozen margarita recipe and one local craft draft beer. The ticket rail never stopped spitting out tickets, bombarding our tiny kitchen with orders. It was a sight to behold.

I found a comfortable corner of the kitchen to work that found me locked in position over the chargrill. I marveled at the crowds descending on us daily as I grilled corn and skirt steak for hours. I've always felt comfortable in busy kitchens, and in some ways, it's easier to hide in plain sight, camouflaged by the activity. Even at six foot four and towering over most of the other cooks, I felt at ease in these first months of craziness. I was in the eye of the hurricane, still and steady.

And that craziness was nonstop and growing by the week. We quickly surpassed our modest financial projections of $1.5 million to $1.75 million annually. In the first few months of operation, it was apparent we would exceed $4 million in sales. After a tough decision to sell back my partnership to Front Burner Group Dining (more on that later), Velvet Taco would peak out at over $5 million—unheard of in this segment. To this day, it's the highest volume restaurant per square foot I have ever been a part of.

Velvet Taco had fast become the darling of our industry; an example of good restaurant site meets good team meets good concept meets good reviews. There were days that we served over one thousand, five hundred people with our modest thirty-seat dining room and another thirty-seat patio that we expanded as the crowds increased. We opened for late-night tacos to serve the after-hours drinking crowd after bars closed, and we opened earlier on weekend days to take advantage of the breakfast taco market.

It was hard work, long hours, loud crowds, tight quarters, and constant lines of customers all the time. The decibel level was sky high. We're in the entertainment industry regardless of whether we acknowledge or deny it. And Velvet Taco was the *Lion King* of taco concepts. The food quality remained high because we had kept the recipes and menu simple. Chef Franke's constant presence and Josh's strong leadership assured us of that.

We'd set out to create a replicable concept with a cool vibe, great food, high volumes, and a strong foundational team to grow with. I started thinking that this would become my next Pei Wei windfall. It felt really good to be on top again. Velvet Taco had seemingly limitless potential and was making money hand over fist for the investors. It lifted me psychologically and potentially financially.

My industry and personal friends started to recognize me again for what I could accomplish. I started feeling as though I'd found a new home. I was hanging out at Velvet Taco most of the day and even some of the late nights so that I could be a part of its nonstop energy and vibe. It was addictive. Chef Franke and I were like mad scientists in a lab filled with possibilities as we worked to create even more edgy and challenging tacos.

We'd created a category we called the WTF taco (a weekly taco feature that was the acronym for What the Fuck), and we gave ourselves no boundaries. We created some excellent tacos that spanned the globe in their flavor and influence, such as the chicken tikka taco—a number one seller. It consisted of fried chicken tenders tossed in a homemade tikka sauce slipped into a warm flour tortilla envelope with buttered basmati rice, cilantro, house-made raita (another Bengal staple), and a sprig of fresh basil.

It was an homage to Bengal Coast and an effort on my part to keep the Bengal Coast spirit alive. It was unheard of to put ingredients like that in that taco. We did the same with other tacos like the Cuban Pig—our riff on "the Oscar" Cuban sandwich—with sliced filet mignon and lump crab in bearnaise and the popular Kung Pao taco—a great wok dish inside a tortilla.

We felt we could charge prices like that of a premium appetizer. So we stretched the WTF to the limits and often created a dynamite ten-dollar taco that no one else was doing or dared do. It was a blast!

But even as Velvet Taco continued to generate attention, praise, and unreal profits, the Front Burner team decided to channel investment dollars into Twin Peaks. That left Velvet Taco in the waiting room. I had come aboard with the idea that Velvet Taco, if successful, would be a growth opportunity, and my experience with other growth concepts would put me in the driver's seat. Randy and I had often discussed that prospect and seemed to agree that it was our mutual goal.

But it was hard to argue that Twin Peaks probably had a higher upside for them in the long run. Velvet Taco, ironically enough, became a back-burner operation for the Front Burner team. It was a downer for me for sure. I wasn't doing much consulting and had planned on converting my ownership to a more permanent position, leading the expansion charge. The future seemed obvious to me, but it wasn't my decision.

The problem was I was never converted to a salaried position. I was removed from my consulting role now that we were in full swing, and my ownership was supposed to be my source of income. This is a very standard arrangement for anyone in a sweat equity role. But while I was indeed considered an owner/partner in the first Velvet Taco, my fully vested ownership wouldn't take hold until I had paid for my share through profit participation. In addition, because I was now the de facto leader of the concept, the expectations fell on me to be the principal operating partner for Velvet Taco. At our weekly meetings, all the operational issues were funneled to me. This made for some pretty contentious conversations.

It was a standoff. I wasn't getting paid, but I was in charge. This wasn't what I had signed up for. Working two years for essentially no pay—with all the operational responsibilities—just didn't add up. Yes, I was an owner but was not fully vested until I had paid back my share of the original development costs—about $200,000. I became disenchanted with the Front Burner team.

I asked Randy for a private meeting one day, and we agreed to meet on neutral ground to talk it through. It was cordial. Randy and I always had an excellent and mutually respectful relationship. But in the end, we found no way to get around this predicament. As hard as it was to do, I suggested to Randy that Front Burner offer to buy me out of my ownership share, and I would move on to other work.

It's important to note that I hadn't financially recovered from my bankruptcy and had all but discontinued my other consulting work as I prepared to grow Velvet Taco. I'd painted myself in a corner and needed resources to cushion me as I got back into consulting. I sat with Scott Gordon to evaluate the value of my ownership.

Unfortunately, Velvet Taco was too young to accurately gauge the worth of my share. And since I hadn't yet begun to pay for my ownership, the present value was nominal. Scott, to his credit, tried to talk me out of selling.

We ultimately settled for $40,000. As I write this, I must pause and shake my head. If I'd been able to hold onto my share and find a way to generate income without jeopardizing my position, I could have cashed out with a far greater windfall. But as pathetic as it sounds, I needed that $40,000. I signed over my interest to the Front Burner team. That decision haunts me to this day. It was one of the worst of my life.

Fast-forward three years. Randy and Jack turned their attention back to Velvet Taco and began to grow it successfully. It drew the attention of private equity firms, which were always on the hunt for concepts with growth potential. The original location was the flagship store that carried the banner for the company.

New locations opened with high volumes but nothing like that first store. The value of the company soared. And the willingness of private equity firms to engage in a bidding war caught industry attention. When the dust settled in 2016, the private equity firm of L. Catterton paid an undisclosed sum for majority ownership of Velvet Taco with just four sites and one under construction.

Typically, PE firms pay from five to ten times earnings for a deal like this. But the undisclosed price exceeded that range—by a lot. The windfall for Front Burner was enormous. After I managed to discover the final purchase price, I calculated the deal would have netted me well over $1 million. But this predicament was one I brought upon myself.

I'm embarrassed to admit it, but a stumble like this is just part of the recovery process. Not all that happened to me after leaving Velvet Taco was bad. I now had another success on my resume, another "Hot Concept" award (my fourth nationally recognized Hot

Concept), and I was back in the limelight. Recognition—almost always my goal in my work—was well documented, and my significant imprint on Velvet Taco's DNA lived on with the concept. To this day, I'm told that the chicken tikka taco—my homage to Bengal Coast—continues to be the number one selling taco throughout the Velvet Taco chain. That alone feels really good.

I am not proud of that diminutive financial gain. But viewing it in a vacuum does not allow for the proper context of what that success meant, nor of what my career—at the ripe old age of sixty— could still become.

And then, one day, a phone call revealed that next opportunity.

10

Trinity Groves. How I Survived Phil Romano. Again

By any definition, Uno Immanivong was a head-turner. As she walked towards our makeshift offices on Broadway Avenue in the rough West Dallas neighborhood, she cut a figure that vivified the eyes. She had long black hair that spilled elegantly over her shoulders, and she wore a tight shirt and a pair of orange slacks that fit as if they had been sprayed on. Her smile beamed from fifty yards away. She was tall—even taller in fashionable Manolo Blahniks— and impossible to look away from.

This was a banker? I asked myself. I was introduced to Uno by fellow restaurateur Mico Rodriguez, an icon in the Dallas restaurant community, whose successes included the powerful Mi Cocina restaurant group. Mico had met Uno through the business and suggested that if I gave her a bit of my time, I wouldn't be disappointed. Her pedigree was exotic. Born in a refugee camp in Thailand after her parents fled Laos, she became a regional sales and support consultant for Wells Fargo Home Mortgage before dabbling in the culinary arts. She had a chef's heart and an entrepreneurial spirit that had impressed Mico.

She dripped confidence as she held her hand out to shake well before I presented my own. If you're pitching an idea to a group of investors and consultants, you probably couldn't have done better. And that's what Trinity Groves was all about: Show up, make a strong pitch, and see if you passed the sniff test and were invited to audition your idea for your first restaurant!

So what was Trinity Groves? Prior to the designation, the area was a destitute, unnamed tract of land west of Dallas's thriving downtown. Years before, it was home to a lead smelter and a cement plant that later became a Superfund cleanup site. Low-income housing, industrial buildings, and a long-haul truck terminal dotted the several hundred-acre landscape. It was accessible for years by the Continental Bridge that linked the area to downtown.

Savvy investors, including my old restaurant buddy Phil Romano, had been buying up land in this area for years, just waiting for Dallas to quench its thirst for growth through westward expansion. It was not a place you would ever visit unless you were headed to Ray's Hardware and Sporting Goods, a haven for gun enthusiasts.

That was before the city of Dallas had enlisted international Spanish architect Santiago Calatrava. Calatrava was to design a series of bridges over the Trinity River as part of the Trinity River Project, a plan aimed at turning the river basin into an ensemble of sports fields, trails, and a nature center and recreational park. One of those visually stunning bridges was to be erected directly next to the Continental Bridge that would be modified into a pedestrian thoroughfare named the Ronald Kirk Bridge.

Financed in part by Dallas's uber-legendary Hunt oil family, the Calatrava bridge was named for heiress and philanthropist Margaret Hunt Hill. In the press and in conversation, the span was sarcastically dubbed "the bridge to nowhere." But never one to pass up an opportunity to seize the moment, Romano, along with his partners, had another idea. Romano spread the word that he was going to convert land at the west mouth of the bridge into a "restaurant incubator theme park."

His objective was to bring dozens of new restaurateur wannabees into a complex morphed from that old long-haul truck terminal, essentially flipping traditional development on its head. Instead of restaurants following residential and commercial development, Romano would use his restaurant theme park to attract developers to the land owned by him and his partners. And attracted they were. They arrived with visions of transforming this nowhere land into the Wall Street or dot-com center of the Southwest, replete with offices, apartments, condos, hotels, and casinos. Even famed luxury casino mogul Steve Wynn (The Mirage, Bellagio) came by for a looksie.

Romano recruited me to help bring his bold restaurant incubator vision to life. Fresh off my success at Velvet Taco, Romano knew that I had the credibility to attract both new and experienced restaurateurs and chefs to this high-profile project—*go west, young chef*. Even with my spotty history with Romano, we made an excellent team, and I enthusiastically signed on. I was the de facto doorman who had to be impressed enough with your idea to invite you to the party and provide an opportunity to open your own restaurant.

I would also be your tour guide, your mentor, and your professor (the nickname Romano gave me), making sure your idea was developed into an executable plan. Trinity Groves would back you by managing your operations and paying you a salary. In exchange, Romano and his partners assumed 50 percent ownership of your concept while charging you rent and a management fee. If your idea was successful, they would be your partners as you expanded, approving your plans while assuming ownership of your brand, splitting the profits fifty-fifty.

The kicker was that the partners would also own 50 percent of the intellectual property rights of each concept. This proved to be a point of contention among some prospects. But while these savvy operators fought back, neophytes were only too happy to have an opportunity to open their own business sans any investment on their part. It was "take it or leave it," and almost everyone took it. Most understood that the worst that could happen was that it didn't work. The upside outweighed the downside.

Uno, with her charismatic approach, was just about as good a poster child for this deal and this theme park project as any of us could have imagined. Never mind that she lacked the experience or that her idea needed some fine-tuning. That was to be expected. She had that indefinable "it" magnetism. Adrian Verdin was the partner in her dream, a pro with restaurant experience and a blueprint to bring their Latin-Asian fusion concept called Chino Chinatown to life. But the deal was made with the understanding that Uno would be the face of Chino. She certainly was.

* * *

Trinity Groves could have just as easily been christened "Petri Dish Groves," given all the experimenting and wild hair ideas we entertained. It was a time when anyone close to the project got a chance to see the absolute best of Romano as well as his bone-chilling worst.

"Can arrogance be so innocent?" I often asked myself.

His partners in the Trinity Groves venture, Stuart Fitts and Larry "Butch" McGregor, were powerless to stop Romano from doing whatever he wanted. That included using Trinity Groves as his personal bully pulpit.

"We're doing something the government can't do," he would say. "We're creating jobs."

The partners even commissioned a filmmaker to develop a reality series titled *The Restaurant Maker*. They paid for three episodes chronicling the dramas at Trinity Groves, not all of which cast them in the best light. Romano was intent on making sure that everyone in the restaurant and political communities knew he was still a major player with influence and a vibrant creative spirit. He was once again puffing his chest, proving to all his adversaries and doubters that he wasn't going away, whether they liked it or not.

The company line was that Trinity Groves was a legacy project for the families of the partners. But anyone who knew Romano and had worked on the project for any length of time was aware that this was much more an in-the-moment endeavor than one calibrated for long-term sustainability. Romano lived for the attention Trinity Groves brought him—from magazine covers to television news coverage—fulfilling what was his main objective.

He had an intense need for the spotlight, which he often generated by being outrageously inappropriate. This was reflected in his paintings: bold, colorful, and full of grandiose strokes and vibrant "look at me" colors. Trinity Groves was his *Mona Lisa*. To see Romano use his roll of onion skin—thin, translucent paper often used for drawing architectural plans—was to observe him in his element. An artist at heart, Romano always viewed himself as better than any designer of restaurants out there. In fact, he would often take a well-thought-out design plan and scrap it because it didn't align with his vision.

Designer after designer came through the offices with creative ideas for a restaurant space that always ended in the same place: the shredder. Other designers who stood their ground were continually second-guessed and cursed by Romano behind closed doors. It was another example of Romano flexing his muscle. With ten to fifteen projects underway at any one time, the situation begged for collaboration and fresh ideas to avoid monotonous design commonality.

Romano micromanaged the process to the point of annoyance. We hired a ravishing-but-inexperienced Italian designer who had not a stitch of experience in the Dallas community. But she was Italian— like Romano—and a rousing distraction for the eyes. Romano put her in charge of keeping his vision aligned with his purpose. I did my best to incorporate her ideas into our plans.

But with the influence and sheer power of will that Romano exercised daily, no one had the cojones to stand up to him. Romano's master plan was to have each restaurant conform to his personal vision. And while no one could argue that he had achieved an impressive portfolio of memorable work over the years, there were whispers that his time in the creative vortex had come and gone.

Just about every restaurateur and chef I reached out to came to hear our pitch and were drawn to the visionary project. The savvier and more experienced of them listened up to the point where they were required to concede 50 percent of their concept to Romano and his investors. Most of the prominent Dallas restauranteurs with iron-clad ideas, like the godfather of Dallas Tex-Mex Mico Rodriguez and restaurant mogul Mike Karns (Meso Maya and Taqueria La Ventana), declined a deal that most assuredly would have handcuffed them to Romano.

There were a few recruits who came to us with novel ideas backed by strong teams to support them. Romano occasionally acquiesced if their pitch was strong and the idea solid. He reveled in his role as the powerbroker. But if the pitched concept came with a design team and some level of credibility, he did step aside. Mostly. His Italian designer was dispatched to keep an eye on any team that didn't want or need his help.

In many ways, his vision was nostalgic as opposed to forward-thinking, more reincarnation than innovation. The first few Trinity Groves projects were indeed homages to his previous work or his childhood memories. Anyone bringing a new idea to the Trinity Groves offices would often be met with blank stares. If something failed to register in his memory bank, he would discount it outright as something that wouldn't appeal to millennials.

If for some reason the rest of us saw merit in a particular project, Romano still would want his fingerprints on it. The beauty of it for him was that if a new restaurant worked, he could boast about all the input he had on the project. If it failed, he could easily dismiss it and point out he had little influence on the idea.

* * *

Romano viewed Trinity Groves as an oil-drilling project. You set up the rigs, drilled, and if you hit a dry hole, you just bore again until you hit a gusher. You could hear him giving interviews that predicted the failure of some of the new concepts, while the rest of us cringed, thinking about what it might mean for concept-and-partner recruitment.

"Hey, come join us," he would say. "And if it doesn't work out, don't worry. We'll just replace you with someone else!"

Now, that's an incentive. What in the world was a man who had accomplished so much doing simultaneously micromanaging ten to fifteen new concepts at the age of seventy-four? Day after day, he'd take some of the wealthiest people from all over on tours of Trinity Groves, often ending at Hofmann Hots, his loud and kitschy New York hot dog stand. It was drawn from another nostalgic memory Romano had during his days growing up in upstate New York.

Over five-dollar hot dogs and '50s and '60s rock and roll blasting on outdoor, unfinished picnic tables, Romano would make his pitch to the moneyed crowd for capital to invest in his development fund. You could see the energy surge through his body as he told his story repeatedly. His intensity was fierce, and he all but bullied you out of interrupting his sales spiel.

In the long run, my best guess is that fewer than 10 percent of the people who heard Romano's pitch invested. Others gauged the risk and ran far away. Yes, the restaurants were going to bring attention to that scrappy acreage. But this wasn't about restaurant incubation in the long run. This was a strong real estate play. If the restaurants brought jobs and attention, other development was sure to follow.

It was a brilliant strategy. Over the course of two years, Romano and his partners paid me a consulting fee of more than $125,000 to help bring restaurant talent to the project. When I started my assignment in early 2012, there were no leases signed, and the land was still strewn with dilapidated, broken-down buildings. But as the PR machine kicked in, and the bridge was finished, there was a constant trickle of interest that turned into fifteen signed leases and ten restaurants opened by the end of 2013.

We'd gotten more proposals than we could absorb. Our main building in that old truck terminal would house up to twelve restaurants at two thousand, five hundred square feet each. Every single one was spoken for! It was a testament to the attention that Romano could command. And even in an unproven area like Trinity Groves, with absolutely zero history, attracting new business was not as difficult as it might have appeared.

In many ways, it was like an arranged marriage. The gist: Take Romano's deal or get the fuck out of the way and let someone else have the opportunity to work with restaurant royalty. The focus was on just how much influence Romano could wield, and how quickly he could become the strong arm of a project. This was evident with the first full-service incubator restaurant.

It was called Kitchen LTO, or Limited Time Only. Casie Caldwell, the partner, was the early darling of the PR push by Romano and his partnership ensemble. Her idea was a restaurant with a rotating theme that would shuffle chefs and interior design on a quarterly basis. LTO was essentially a permanent pop-up restaurant—an incubator within an incubator.

Casie was an accomplished restaurateur with a keen sense of marketing and social media. Her idea was well pitched, her experience strong. She brought an engaging personality and a level of solid professionalism that was always on target. She became the face of our early development. But as the concept started to take shape, it was clear that Romano had fallen out of love with the idea— not because it was bad, but because Casie seemed uninterested in Romano's input. She had chosen a respectable designer for the space, and as their design came into focus, Romano grew combative. As more restaurant ideas were presented to him, Casie and her concept lost their luster. But she was persistent, and she doggedly pursued her dream, while Romano privately schemed to stall construction so that other restaurants could open before hers.

He abhorred the idea that she would be first out of the chute. Shortly before completion, he floated the idea of not opening Kitchen LTO at all and just letting the space lay dormant. His partners convinced him that this would be a PR disaster. So, after tense negotiations, the project continued under Romano's intense scrutiny.

As the days dwindled down to the opening, Romano's mood further soured. There was anger rumbling through his footsteps, and his face turned a ghastly shade of red as he stormed into the almost-finished Kitchen LTO. There was no masking the volatility in the air as two distinctly different points of view were about to collide. Casie was enthusiastic about the restaurant design, while Romano was intent on emphatically laying down his rules on restaurant development for future partners.

Expletives flew out of his mouth like a torrent of bats breaking free from a cave as he raged his contempt for the design. Casie melted under the brutal harshness of Romano's fury, thickening the room with tension. This was a turning point for Kitchen LTO: Either adapt to Phil's vision, or the restaurant would never open. I did my best to intercede. I knew how vital this restaurant opening was to Trinity Groves, and I was a big fan of Casie's work and the passion driving her vision.

Because of my experience with Romano and his incendiary methods, I knew that after he aggressively laid down his law, he would be satisfied knowing he'd intimidated everyone in the room and established his alpha-dog superiority. At that point, there's an opening for reason and negotiation.

"Phil, let me work with Casie and her team and see how quickly we can get this done," I suggested. "Give me a few days, please. I'll take care of it."

He lowered his voice and began pointing out what he saw as the design flaws he wanted to be fixed. I made notes. My goal? Get the restaurant open and keep Casie focused on the food, the staff, and the opportunity. If Romano had to have his way with the design, it would ultimately be a small price to pay if the restaurant could open on time and build momentum. Romano strode off, and I calmly began my work with a very rattled Casie.

Color schemes were hastily changed, décor ideas were scrapped, and egos were bruised. But in a few days, Romano reluctantly awarded his approval, and we pushed forward. At the opening, the food was excellent, and opening chef Norman Grimm performed as well as could be expected under the circumstances. Sales were predictably light since most of the construction at the complex was still underway. The entire building looked like a rubble field borne of an artillery barrage.

But initial reviews were positive, and LTO survived the choppy waters all new restaurants endure. Trinity Groves was now on the map with a success. Casie successfully developed her business and somehow reached profitability levels that took her off Romano's target list. As other restaurants and shops opened over the ensuing months, Casie relaxed, and her spirit improved.

Little did she know that Romano had secretly met with other chefs and restaurateurs to gauge their interest in taking over the space in the event it didn't work out. Romano was a fucking pit bull with a long memory for those who go against him and an insatiable hunger for winning.

* * *

It was the summer of 2013, and very little was known about our incubator on the other side of the bridge. To address this lack of limelight focus, I stole a page from the cable food networks and their slate of competitive cooking shows. I would pit our newly signed chef entrepreneurs against each other in a series of cooking competitions I dubbed "Fork Fight at Trinity Groves."

It was a March Madness-style bracket contest, where winners advanced and ultimately earned the opportunity to face off against our biggest hitter: locally acclaimed chef Sharon Van Meter. Van Meter had a large event hall at Trinity Groves and did catering out of a full kitchen she had built in one of the larger buildings. We promoted the living hell out of the series, selling individual tickets to the dinners. They featured four courses prepared by the two competing chefs for seventy-five dollars a person, with an open bar donated by some of our vendors looking for long-term relationships.

Over the course of a steamy Dallas summer, we held dinners in the event hall in sweltering heat with a willing and enthusiastic sell-out crowd of three hundred guests for each event. They voted for their favorite with forks we provided, dropping them into one of two wine buckets we placed on a table—one for each competing chef. The bucket containing the most forks at the end of the night was declared the winner.

These events got tons of local media coverage and generated attention for Trinity Groves restaurant openings that exceeded our expectations. The eventual grand champion of Fork Fight? That charismatic neophyte restaurateur/chef and former banker Uno Immanivong after winning three rounds. She had gathered a small army of supporters to help her prep her southeast Asian cuisine. Her Laotian mother, who spoke little English, was the inspiration behind Uno's cuisine and was always watching over the preparation and the final plating as the dishes left the kitchen.

My fellow restaurateur Bob Sambol, the founder of the famed and lucrative Bob's Steak and Chop House, was right by my side as we oversaw the show. Sambol had joined the Trinity Groves team initially as a strategist for Romano's Hofmann Hots concept. His experience was instrumental in guiding recruits from the wide-eyed stage to the next level of sober restaurant management.

When the dust settled, everyone at Trinity Groves was a winner. We were no longer a plot of barren land on the other side of the bridge. Trinity Groves was now well known throughout Dallas. I'd gotten closer to Uno during this period of six months, and though I continued to harbor romantic feelings for her, I didn't want others at Trinity Groves to think she'd gotten her opportunity because of my interest in her. Her victory at Fork Fight helped dispel lingering rumors that she was my pet project.

I often wince when I think of my awkwardness dealing with my attraction to her. Once, during lunch at a local Mexican restaurant, I professed my fondness for her and just as quickly declared that I wouldn't act on it because of the possible professional damage it could inflict on the both of us. I sensed she had the same kind of attraction.

Call me tone-deaf. It wouldn't be the first time. But at that same lunch, I asked her if she'd let her stunning sister Liz know I was interested in dating her as well. My reasoning was that it would keep me close to both if anything changed. Liz and I did go out to dinner once, but it never went beyond flirting, and neither relationship developed into anything. I stumbled all the way through.

I regret not handling the situation better, but I remained on good terms with Uno and watched with pride as she opened Chino to capacity seatings and a solid reputation. Bob ended up being more of a mentor to Uno. But in the long run, the chef/restaurateur bloom was off the rose for her, and the romanticism—with the relentless energy and time a new restaurant requires—wore off.

On top of all those demands, Uno was a single mother to a young daughter who needed more of her time, energy, and guidance. And if that wasn't enough, when you have Phil Romano as your partner, there's never a release valve for the pressure he either intentionally or unintentionally applies. Bob did his best to act as a buffer between Romano's often blunt and harsh criticism as well as his sometimes-off-the-cuff remarks. But as accomplished as Bob was, this became a lot like swimming against a strong current that keeps you treading in place.

Romano's style intimidated all the incubator operator-partners. I didn't envy those like Uno who simply weren't ready for that level of intensity. Chino Chinatown lasted about four years, the last of which Uno had no role in the operation. Eventually, it became another dry well as Romano took over the space to open—I kid you not—a Chinese restaurant called Sum Dang Good Chinese.

That restaurant is still operating as I write this, despite complaints from the Asian restaurant community that the name and menu references to Chinese culture were insulting. Menu blurbs included doozies like "Who Doesn't Like a Happy Ending?" over the dessert section. These attempts at cuteness drew passionate responses from local chefs and restaurateurs, including one targeting Romano's alleged sense of humor.

"Grow the FUCK up and ACT like a professional business-person with some sense and human decency," commented well-respected Asian chef Reyna Duong on Twitter.

Romano brushed off these complaints as politically correct bull, but he eventually and reluctantly acquiesced, removing the offending verbiage.

* * *

Trinity Groves went on to become a huge hit over a short period of time, attracting people from all over the Dallas-Fort Worth metro area. It took little more than a year for the novelty to wear off and reality to set in. Well-heeled diners saw no point skipping the fine fare and environs served in Dallas's Uptown and downtown restaurants in favor of Trinity Groves's over-the-bridge eats.

And while there were many chefs who drew praise for their efforts, most of the first-timers just didn't have the necessary stamina, vision, or team cohesion to sustain long-term success. Nor did they have the energy to fight Romano's ever-present oversight.

Trinity Groves served some excellent cuisine, such as Casa Rubia, with up-and-coming chef Omar Flores; Amberjax, a Louisiana-style seafood shop; and Cake Bar by one of my favorite entrepreneurs, Tracy German. Tracy worked her butt off to build a reputation for baking the best cakes ever served. She hit the mark. Finally, chocolatier Kate Weiser established a brand that flourishes to this day.

In fact, Tracy and Kate were the only operators ever to have expanded beyond their Trinity Groves confines. Tracy added a cake production facility in Dallas. And Kate opened locations in the famed upscale NorthPark Center in Dallas and in Fort Worth in addition to a production facility in the nearby town of Garland.

The Cake Bar and Kate Weiser Chocolate are the only two remaining of the original incubated twelve still owned and managed by their creators. Not a great batting average, especially once you consider Romano's own concepts—Hofmann Hots and Potato Flats, a baked potato emporium—closed quickly and quietly.

The story of Trinity Groves is that it created far more opportunity for Romano and his partners than it did for these budding restaurateurs and the investors who sunk cash into their restaurant investment fund. Yet they never achieved the development luster they promised. Besides apartments and venues like Lone Star Axe Throwing, none of the visionary Wall Street, dot-com, and casino/hotel-style development ever emerged on this scrubland. In the end, Trinity Groves is just another mediocre place to dine and imbibe.

When I look back on the two years I spent with Romano on the Trinity Groves project, it is with a mixture of sadness, anger, and disappointment. I spent a great deal of time drinking Romano's Kool-Aid and genuinely believed the original vision was rich with opportunity. I recruited many chefs, restaurateurs, and dreamers into that flight of fancy and worked diligently for their success. The Fork Fight series, the time I spent teaching classes ranging from Daily Decision-Making to Restaurant P&L 101, and the time spent assisting with openings drained me of far more hours than I was compensated for. My own entrepreneurial spirit is embedded in those buildings.

We're all dreamers, and I'd never discourage that. I know intimately that failure is a part of the journey. But so many good people just didn't enjoy the rich opportunity they were promised, and for that, I hold Romano and Stuart responsible. Few who participated in those early years would disagree with that sentiment.

Romano offered me the option to stay on at Trinity Groves and transition into a management company to supervise the new restaurant and food businesses Trinity Groves opened. But I was tired of the hectic, unpredictable atmosphere and Romano's histrionics. We had a short conversation, and he was convinced I turned down his offer because I was an underachiever who never really capitalized on my innate talents the way he had.

"Brezinski, you seem to have always resented me because I've achieved at this level, and you have only ever achieved at this level," he said, positioning his hands to demonstrate the vast success gap between us.

I didn't respond. The silence was a fitting end to our collaboration. I would occasionally return to visit my friends and sit with Bob Sambol to get updates. Bob had accepted Romano's offer to create that management company and did well for several years. But he, too, reached the end of his patience with Romano—I call it shelf life. He returned to Bob's Steak and Chop House to rekindle and elevate the success of the concept he founded.

We all wake up eventually.

11

Banh Shop. My Fingerprint on Yum!

I was experiencing another lull in my work, scrambling to fill in the time after I prematurely sold my share in Velvet Taco. I needed money to pay back friends, pay bills, and cover debts and back taxes I'd accumulated while recovering from bankruptcy. The consulting work I was getting was sparse, and I continued to try to find ways to cover my monthly expenses.

One day, in 2013, when I still had a telephone answering machine, I came home to find a confusing message from someone with a French accent. The callback number started with 502, which, from my travels, I recognized as a Louisville, Kentucky area code. I'd opened a Macaroni Grill there about two decades earlier and recalled the area code. But I could not make out the message.

Half of me wanted to erase the message—you do that a lot when you've been barraged by creditors looking for payments. The other half was intrigued. After a day or so, I finally called back. It was one of the better decisions I've ever made.

"Hello," said the heavily French-accented voice on the other end of the line.

"Hi, it's Mark Brezinski returning your call, but honestly, I could not make out the message," I said. "So you'll have to start from the beginning."

The voice belonged to Christophe Poirier, an executive at the largest restaurant company in the world: Yum! Brands. Yum! Brands owns Pizza Hut, Taco Bell, Kentucky Fried Chicken (KFC), and, years later, The Habit Burger Grill. Christophe worked at the new Yum! corporate headquarters in Plano, Texas. During our short conversation, he explained that he had moved to Dallas and had been put in charge of a division to develop new Yum! Brands franchisable concepts. (Yum! franchises some 94 percent of their restaurants.) That division was dubbed Yum! Emerging Brands.

Christophe had read about me after coming across Velvet Taco while surfing the net. He was intrigued by an article written by a food critic who ripped Velvet Taco for serving "fake food." I had written a diplomatic response to the reviewer's criticisms to avoid a social media brawl, which prompted the writer to publish my response. Christophe had read the trail of communiqués online and was impressed with how I had handled the situation. I'd kept a level head, didn't get defensive, and offered to host the critic at Velvet Taco to show him how we composed our menu offerings and to open our operations for his inspection. You can read the whole thing if you Google Velvet Taco–Taco Trail–Mark Brezinski. The reviewer "Taco Trail Jose" (real name Ralat Maldonado) had taken me up on my offer. The ensuing online review smoothed the water a bit, and the social media frenzy died down.

In addition, Christophe worked with in-house Yum! Brands designer Steve Chalsson. Steve had become a regular at Bistro Babusan, a fresh Asian fusion concept I created north of Dallas in the town of Fairview. He loved what we had done with the space and the variety of Asian cuisine we served. He shared his positive impressions of Bistro Babusan with Christophe, further cementing my opportunity to work with him on new projects.

I met Christophe the next morning at the Original Pancake House in Addison, Texas. We met at nine-thirty in the morning and talked straight through lunch. Christophe (hereon CP, which is what I ended up calling him) must have liked what I had to say. And, as luck would have it, he had a passion for banh mi, the famed Vietnamese sandwich, and wanted to develop a concept to feature it.

Around this time, famed chef Anthony Bourdain declared the banh mi he'd eaten in Vietnam during his travels for the television show *No Reservations* to be the best sandwich he'd ever eaten, calling it a "symphony in a sandwich." Turns out I had done quite a bit of research on banh mi during my Pei Wei years and had also developed a fantastic version of it with Chef Nevielle Panthaky for Bistro Babusan.

CP was thrilled I had a history with the sandwich and asked how quickly I could put together a proposal for a concept using banh mi as the focus. He was clear that this was a front-burner assignment and that Yum! was prepared to put a considerable amount of money behind the development of this and other concepts.

There's a stark difference between developing your own restaurant project, like Tin Star, and putting together a new concept for a team, like Yum! Brands. At Tin Star, we cobbled together a few hundred thousand dollars and spent many, many unpaid hours developing recipes. At Yum! Brands, we had millions of dollars, full corporate kitchen facilities, and topflight marketing, graphics, and research services at our disposal. Even before we started construction on the first of our new banh mi-centric restaurants, the spend was well into the million-dollar-plus range. We racked up expenses in travel, game-planning, food testing, consultants, and the salaries of Yum! Brands people assigned to our Emerging Brands division. It's all relative, right? Tin Star—from soup to nuts—cost about

$700,000 from our original thoughts to opening, with no less time, effort, and research.

Once CP accepted my proposal—one that would free me from my financial woes for several years to come—we set about developing strategies for this yet-to-be-named banh mi concept. With Yum! Brands' war chest at my disposal, I tapped Yasmin and Braden Wages for this project, a highly respected, food-knowledgeable husband-and-wife team.

They'd both attended my alma mater, Cornell University's School of Hotel Administration, and had deep training with the country's most acclaimed restaurant company, Hillstone (at that time still called Houston's). They owned and operated a great little place in Dallas called Malai Kitchen, which featured some of the same culinary focus I'd served at Bengal Coast but with a broader Southeast Asian influence and an emphasis on Vietnamese flavors.

The Malaysian peninsula in Southeast Asia is home to some of the brightest and most alluring cuisines in all of Asia. Yasmin and Braden were well versed in replicating these great flavors and dishes. It didn't hurt that they were young and energetic, bursting with charisma, and had extensive dining room and kitchen management experience. They'd also traveled frequently to Vietnam and had done tons more research than I had. I was glad to play second fiddle in their culinary contribution to this new project.

They inked their own agreement with Yum! with my guidance, and we set out to develop a new concept that would break the mold of the more familiar foods of Yum!'s core brands. I didn't yet have access to the upper-level Yum! executive team, but my guess was that they realized food trends were expanding beyond their current portfolio, and they needed to step up their game to compete in this quickly emerging Asian segment.

* * *

My first brush with this giant restaurant monster called Yum! happened years before. While CP discovered me through his own research, I'd first been introduced to the company by Micky Pant, a true gentleman who was, at the time, the global president of Kentucky Fried Chicken. When I owned Bengal Coast and was struggling to make a go of it, Micky had quietly become one of my most ardent supporters, unbeknownst to me.

Micky is of Indian descent and had found my Bengal Coast through articles written about my risky venture in a handful of papers and magazines. The Indian community, for the most part, was extremely supportive of the restaurant and is probably the reason I survived as long as I did. One evening at Bengal Coast, Micky approached me and introduced himself.

He asked me if I'd be willing to speak to an International Conference of Kentucky Fried Chicken franchisees from around the world who would be coming to Dallas in a few months. I agreed, partly because I wanted to please Micky and partly because I thought the exposure might be good for Bengal Coast. For years, my favorite fast food had been the original recipe KFC chicken thighs. I knew that if I shared a short story about my obsession with these seasoned chicken parts, it would kick my presentation off on the right foot. I rarely passed up an opportunity to ham it up in front of a crowd, so I prepared for the speech.

A year before my date at that conference, I had been visiting my sister in the Cayman Islands. She is an artist and had a small studio condo along Seven Mile Beach—always a great getaway destination for me. On one of those trips, I experienced late-afternoon hunger pangs, and as it turned out, there was a small KFC location in the small strip mall across from her condo. I hiked across the street and indulged in three original recipe chicken thighs. That snack hit the spot, as it always had for me.

Later that evening, my sister and a few of her friends took me out to an acclaimed and expensive restaurant on the west end of Grand Cayman. I explained to that audience of franchisees that the KFC chicken I had earlier in the day was much better than the expensive meal I had at that four-star restaurant.

The crowd loved my story, and the room erupted with applause and laughter. I'm not even sure what I spoke about after that, but I had won over the room and earned their respect. As the evening wound down, one of the franchisees in the audience approached me with a proposal to work with them in the United Kingdom. That conversation ended up kicking off my work with Pizza Hut UK, as I shared in the Aftermath chapter earlier. If it hadn't been for Micky Pant and his belief in me, I might not have ever gotten the Yum! Brands work that I enjoyed over the years.

But back to the emerging brands concept. For about six months, Yasmin, Braden, and I were inseparable. We worked on menu ideas, held tastings for Yum! executives, and had carte blanche to create whatever food we felt was right for this banh mi-centric concept. The tastings, decision-making, and ability to perform under the lights came naturally to all of us.

As we debated name choices, I commented to CP that we didn't need a cute name and that it should be simple. I reminded him that all Yum! Brands' names embodied the type of food they were serving—Pizza Hut, Taco Bell, Kentucky Fried Chicken or KFC, and, years later, The Habit Burger Grill.

"What about something as easy as Banh Shop?" I asked.

From that moment on, we had a name and a focus for the rest of the branding. We also found a location that, while not exactly at Main and Main, was an excellent spot. It was a free-standing building on SMU Boulevard housing an old machine shop near Southern

Methodist University, just north of downtown Dallas not far from the original Velvet Taco.

We made some great collaborative decisions along the way, but our team tripped up by adding a big red star to the logo. To us, the star signified strength, and the signage was impressive, similar to the Macy's red star used in their logo as an apostrophe.

I'm embarrassed to admit we didn't realize the negative implications of that star due to its association with the years of communist rule in Vietnam. After a social media backlash, we pulled the sign, developed a logo sans the red star, and got Banh Shop open in September 2014. We quickly and successfully averted an extended online lashing over the blunder. CP was beaming with pride. The food was outstanding, and Bahn Shop's Saigon street-food vibe was authentic. That Josh Garcia, the general manager of Velvet Taco, resigned and came on board was the icing on that cake.

"It's remarkable what you can do when you take a small team of good people and leave them alone to create something like this," said David Novak. Novak was the CEO of Yum! Brands, and he shared this comment with me during a private Banh Shop tasting with the executive team. No doubt, everyone was ready to unveil this new vision to the public.

Yet despite all that positivity, Banh Shop struggled. Admittedly, the location wasn't prime, but our neighbors at the time included very successful fast-casual restaurants such as Torchy's Tacos and Twisted Root Hamburgers. Our access was easy enough, the demographics matched our profile, and our team was executing perfectly. But we were fighting an uphill battle gaining a foothold in a highly competitive market.

In other words, consumers develop well-studied eating-out habits that are difficult to break. New concepts, like Banh Shop, need to focus on ways to break those patterns and persuade consumers to adopt new dining routines. We call this "breaking into the rotation."

Not easy. We knew the eyes of Yum! executives were watching closely, but CP offered unconditional support. We believed we just needed more time to get into the food rotation of our area.

But one morning, a huge pin pricked our balloon. Workers and trucks from the City of Dallas showed up on our street directly in front of Banh Shop. *Harmless enough*, we thought. Maybe they were going to do some minor street repairs or utilities work. Not so lucky. They were soon reinforced with even more trucks and workers, and they began tearing up the entire length of the street.

They occupied the street for more than six months, often completely closing it off to traffic. Our parking lot—a huge advantage in an area that had mostly street parking—was intermittently blocked as new sewer lines were installed. It was a nightmare, deadly for a new business. We had no recourse.

Our Yum! support waned. To compound matters, David Novak was retiring, and the new CEO, Greg Creed, was quick to mention that under his leadership, Yum! would focus on its core brands. The die was cast. We scrambled to come up with a plan, which included trimming expenses. I wasn't being paid, but we still had Yasmin and Braden on the payroll.

That had to end. We had no choice. CP called an emergency team meeting at Yum! headquarters. It was an uncomfortable meeting, and it ended with hurt feelings and much disappointment. After more than a year of being tied at the hip, preparing lunches and dinners, testing food, and developing a team, it was over.

* * *

One of the regrets I have about this project was that that meeting ended my strong relationship with Yasmin and Braden. While I was used to a hard stop with consulting projects, I think Yasmin and Braden saw Banh Shop as an opportunity to continue to earn a handsome income. But the resources just weren't there. They thought I had conspired with Christophe to cut them out. But that simply wasn't the case. My own fees had been stopped months before, and I'm not sure they ever knew that.

Yet there was a glimmer of hope. Shortly after Banh Shop opened near SMU, we had an opportunity to convert an under-performing concessions space at Dallas-Fort Worth International Airport's Terminal D into a Banh Shop. It was wildly successful, far more so than the streetside location. Despite this, Yum! was ready to jettison the concept and shut it down. A company with forty thousand restaurants worldwide doesn't have the patience to feed a new concept without seeing solid potential for returns. Banh Shop was going to have to make it on its own or not at all.

But I wasn't ready for a decision of that severity. My stubborn belief was that Banh Shop had a bright future. I successfully convinced CP to plead for more time. I offered a plan that provided Yum! with a way out and helped the company recapture some of their initial investment while retaining a minority share if we expanded Banh Shop successfully on our own. It took a while, but I found a buyer, and we negotiated a deal to purchase 80 percent of the concept for a nominal price, giving Yum! a 20 percent retained interest.

In addition, we gave Yum! the right of first refusal to purchase the concept back from us if we ever grew it to a reasonable number of locations. It was as much of a win-win as you could hope for. We had both ends of the spectrum: an insanely successful airport location and a deadbeat streetside location. In the Yum! world, it was hardly good enough. But in our world, it was a lifeline.

Our savior? Derek Missimo and his D&B Mitchell Group. Derek and his team had helped us convert their struggling airport location into Banh Shop. They were ecstatic with the results. Sales were teeming, and the lines to get banh mi and our Asian wok'd bowls and fresh spring rolls made it one of Terminal D's busiest spots. We won national awards as one of the best new airport concepts, and Banh Shop started generating more attention.

Our newly formed company, YEB II, moved forward with determination. We quickly set about developing a plan for Banh Shop's potential growth. Derek and I complemented each other well, and with Yum! as a silent partner, we could continue to promote it using Yum! logos and presence. We closed the SMU location and abandoned the idea of trying to create more streetside units, focusing instead on new airport opportunities.

We found a new tenant for the SMU Banh Shop, effectively taking Yum! off the hook for the lease. Derek, in addition to his other skills, dedicated his energy to promoting Banh Shop with his friends at SSP Group—an airport concessions firm. SSP was enamored with the volumes that Banh Shop was generating at the small space at DFW International and helped us ink deals in both Toronto Pearson and Vancouver International airports. Soon, Banh Shop was operating in three international airports as a successful template was being developed.

Banh Shop had new life and continued to flourish as an up-and-coming airport concept. Toronto's version opened with huge numbers, and, while not as robust, Vancouver was also doing well. During this transition, I was no longer making conceptual decisions. Nor was I being paid by Yum! I had a nice ownership piece of the newly created company, but the future was now in the very capable hands of Derek and his team. I moved from the decision-maker to a more silent partner. Time to find my next project!

But that wasn't going to be easy. For about two years, I'd completely immersed myself in Banh Shop and had no other clients. And, as always, the bills didn't stop. So new projects had to be scared up.

As of this writing, I still have a piece of Banh Shop. But with the onset of COVID-19 and the malaise that hit the travel business in 2020 and 2021, expansion ground to a halt. There are now four

Banh Shops operating, and I feel in my heart that there is a great future for the concept. My faith in both SSP and Derek remains unshaken. My work with fast-casual concepts and my experience with Yum! bolstered my confidence and view of myself as an innovative concept creator.

My reputation within the Yum! corporate offices carried over to my next opportunity. Soon after leaving day-to-day Banh Shop operations, I received a call from Anne Fuller, a Yum! executive. She'd gotten my name from CP and wanted to know what I knew about pizza, particularly the popular new wave of "build your own" (BYO) pizza concepts that were disrupting the market and creeping into Pizza Hut's piece of the pie. Yum! was ready to step in and protect their turf, and I was asked if I was able to join a new team dedicated to that purpose. Here we go again! I jumped on board for the ride.

12

Cashing in on New Dough

Chef Enzo Minopoli, Yum! Brands executive Anne Fuller, and I were breezing down Highway A1 from Rome to Naples in a rental car with the windows wide open. It was an idyllic scene, the cool air breezing over three people driving through the Italian countryside without a care in the world. We'd just spent time in Florence, drinking Vermentino at a piazza café, dined al fresco in Tuscany at a small vineyard, and enjoyed a week of research for Yum! Brands' newest entry into the global pizza wars.

Rome, Naples, Florence, and Tuscany, with stops in Sarno (the breathtaking home of tangy San Marzano tomatoes) and Alvignano (the awe-inspiring home of creamy buffalo mozzarella). Our small team, with oversight from Christophe Poirier back in Dallas, was given the mission to help Yum! Brands join the BYO (build your own) pizza market, the craze sweeping the US. At that time, in 2015, new BYO pizza brands seemed to pop up weekly.

Blaze, Pizza Rev, Pie Five, Pieology, and several other regional marques were gobbling up market share and threatening Pizza Hut's stranglehold on the eat-at-home pizza segment. As with many dining trends, BYO pizza had its genesis in California. We were striving to determine how to enter this rapidly growing segment with solid differentiators that set us apart.

I reveled in this work. My experiences with Yum! Brands proved to be the prescription I needed to recover from a dark period in my life while providing the security of both purpose and income. The Yum! family inspired me to bring my best game to the table each time they requested my assistance. I was beyond the age of having to prove myself, but with lingering insecurity over my failures continually gnawing at me, Yum! was a godsend, and every assignment became an obsession.

Chef Enzo was a native of Naples, and our stop there included a dinner at his parents' home. But that would have to wait. After we'd arrived at our hotel in the heart of Naples, we immediately hit the narrow streets in search of the legendary Neapolitan pizza. We hadn't yet determined what, if any, role this style would play in our new project, but no tour of Italy and no study of pizza is complete without an exploration of the delectable Neapolitan style.

Differentiated by its chewy, soft, wet, and thin crust, along with the simplicity of its toppings, Neapolitan-style pizza is famously cooked in specially constructed wood-burning ovens at seven hundred degrees and higher. Slightly larger than individual-sized pizzas, these gems are cooked in under two minutes, are never cut with a roller, and are always eaten with a knife and fork. No apologies and no exceptions.

Several pizzerias were on our hit list, and we wasted no time following Enzo through the claustrophobic and ancient cobble-stoned streets to our two main targets: Starita and L'Antica Pizzeria de Michele. We were in the belly of the Neapolitan pizza beast, and time was wasting. Our dinner at Enzo's childhood home in the residential hills was important for our team, but our main reason for being in Naples was pizza, and we indulged.

It's so easy to get caught up in the romance of Neapolitan pizza. Simple, authentically local toppings; crust that blisters with a slightly salty and fresh-baked aroma, and the softness of the chew bursting with tomato flavor, garlic, and the creamiest mozzarella I'd ever tasted. But romance is one thing; business is another. And we were there on business.

We knew several Neapolitan pizzerias dotted the US landscape, but all were regional players. Cane Rosso in my hometown of Dallas; Antica Pizzeria in Atlanta; 800 Degrees Woodfired Kitchen and Pitfire in LA; and Don Antonio by Starita and Keste in NYC. Even Chipotle had gotten in the game by purchasing a small pizzeria in Boulder called Pizzeria Locale. They had built a new spot in Denver on the way to a rapid expansion to go national that would ultimately fail.

No arguing the excellence of the product, but Americans en masse like holding their pizza slices, not knife-and-forking them. Neapolitan authentic? Absolutely. Convenient? Not so much. Pizza on the run, by the slice, or cut into eighths by your local pizzeria is an American tradition. Our determination after this glorious stop in Naples was that this style would not be in the running for whatever we created for Yum! Brands.

I imagined upon hearing this that the entire population of Naples would rejoice! A slice of pizza in Naples was as rare as fresh oysters on the half shell in Des Moines. We left Naples focused on another style of pizza we'd eaten in Rome, a creation called Pinsa Romana. When in Rome, we had heard about a special type of flour—pinsa— that a family was producing just outside of this ancient city.

The brainchild of the Di Marco family, this remarkable flour was a combination of high-protein 00 wheat flour and two non-gluten flours: soy and rice. It produced a crust that was light, airy, and soft on the inside, with large air pockets like those found in ciabatta bread. The outside was crunchy but delicately so sans the brittleness of cracker crusts. Several pizzerias were serving this wonderful pinsa flour, so our tour included as many places as we could fit in.

The most memorable of those was a small pizzeria in the shadow of Vatican City: Bonci. Locally famous chef Gabrielle Bonci had concocted his own version of pinsa flour using a formula that contained ancient grains but ended up having the exact same texture as pinsa romana. Bonci's pizzas were fabulous! Our cameras clicked, and our stomachs were jubilant.

To this day, the food photos I took at Bonci are among the best of my years spent exploring foods from around the world. So simple, so creative. Bonci's display cases (his pizzas were cooked in large rectangular pans, cut to order with scissors, and weighed on a scale) were addictively attractive and impossible to look away from.

* * *

With the Italy trip now in our rearview mirror, Anne, Enzo, and I set out to assemble our game-changing proposal to Yum! Back in the States, we commandeered a bright Yum! researcher, Brock Johnson, to help us with our daunting assignment. I also reached out to Chef Mark Miller. We needed a flavor expert on our team to help Enzo acclimate his style and palate to the American consumer. Mark's addition and experience made us an even more formidable team.

What we had learned was that Blaze had proven that the US consumer was open to individual-sized pizzas with custom toppings cooked quickly in super-hot ovens and served on traditional round aluminum pans. Blaze highlighted the show business side of pizza preparation in a fast-casual environment with a queue line common in other fast-casual concepts (Pei Wei, Panera, etc.). Blaze also offered easily selected choices from a toppings bar display and ovens managed by flamboyant "pizzaiolos" (the term for a person cooking and "spinning" the pizzas in the ovens). Because of their smaller pizzas, the thinness of the crust, and the temperature of the ovens, these individual pizzas cook in under ten minutes—an ideal time frame for fast-casual diners.

There comes a time in all conceptual development when the creative team must face a question: "How authentic do we want to be?" With pizza, authenticity is mostly about the crust, along with a few expected toppings. Blaze and all the others didn't diverge from this protocol. They just committed to doing it faster and in personalized sizes while composing the pies right in front of customers at a digestible price.

Those of you from New York City might scoff at this because you've been eating pizza by the slice since you were knee-high to a grasshopper. You're used to seeing up to a dozen whole pizzas in a case and the pizzaiolo using a rolling cutter to carve off a slice and warm it in a deck oven. You're getting exactly what you want just about any time you want it. So this Blaze-craze makes little sense to you.

But this is rare in more than half the country. In Chicago, deep-dish pizza is the rage: a pie that takes up to forty-five minutes to bake. The same is true with Detroit-style: a thicker, doughier pie that doesn't translate well in the Southwest and beyond. Our sweet spot in the pizza wars was to compete with individual-sized pizzas but with one major point of difference: the crust. And because of this determination, our study went back to Rome and the special pinsa romana.

We held tastings and dough-making sessions. We proposed collaborating with this small flour factory to provide us with exclusive access to this unique flour—a long-term partnership to introduce pinsa romana to the world. We were well ahead of anyone in the quest to introduce the next best thing in pizza.

We were onto something, and the energy was palpable as we introduced more and more people to this new style of pizza. Our next step was to develop a menu, come up with a catchy name, and create a prototype. These projects can cost millions of dollars in research and development, and it was clear that Yum! was not sparing any expense because of the potential impact of this new brand.

We traveled to London to hear pitches on marketing work, then back to Rome to eat more pinsa romana, and we game-planned how to put together a tasting for the Yum! executive team in Dallas. Our culinary team—magicians with this dough—was creating some of the best flavors I've ever tasted. Over a period of months, we tested a multitude of ovens, experimented with using the pinsa dough as a sandwich base, tested salad recipes, and put together a menu that was both unique and executable at scale. We began planning the all-important executive and franchise tasting that would essentially determine the future of the concept.

Anyone we informally invited into the kitchen to taste our versions of BYO pizza on pinsa flour gushed with praise. This mouthwatering pizza wasn't too heavy, too chewy, artificially enhanced, or drowned in seasonings. Its simplicity was a revelation to just about all of us. We'd managed to recreate Neapolitan-style pizzas with pinsa romana. We unveiled this new concept with a catchy name: Zefiro.

The graphics were bold and precise. The overall presentation was smart and contemporary. The name was easy to say and musical to hear. The hurdle? Yum! Brands is almost exclusively a franchise company. Just a tiny fraction of its restaurants is company-owned and operated. Zefiro was being developed as a franchise vehicle, one with no sales history. We had to find a franchisee or two within the Yum! world willing to risk betting on an untested concept and build prototypes to prove it in their markets. Tough ask.

* * *

Franchisees typically do not develop concepts. Instead, they provide expertise in replicating proven models and operate them to their limits. Yum! has over two thousand franchise partners, and their franchisee network is second to none. In addition to inviting a hand-selected group of US-based franchisees, this tasting had added the pressure of being staged in Plano, right in the shadow of the Yum! executive team.

The day of this grand tasting will live in my mind forever. There was a hand-selected group of US franchisees, dozens of Yum! senior executives, and the weight of the possibility of creating a new global brand. We had command of the entire Yum! test kitchen, a sprawling space larger than most hotel kitchens I'd been in, and assistance from most of its staff.

There must have been one hundred people jammed into the space, double or triple any tasting we'd held previously and about ten times as many as I'd ever had at a tasting before. Standing room only. The buzz grew louder as the anticipation peaked. Our team was prepared to present six variations of our individual-sized pinsa romana pizzas. Toppings ranged from tomato sauce and cheese to a standard pepperoni pizza to our team's favorite: BBQ chicken and sausage pizza with red onions and cilantro.

I was sure we were going to oversubscribe our limit on large franchisee commitments. The collective net worth of the individuals rubbing elbows in that kitchen would have been mind-blowing to calculate. A staff of servers wormed through the crowd, filling beverage orders and adding a sense of hospitality to the moment.

Looking over my shoulder, I could see Enzo, Chef Miller, and Brock skillfully orchestrating the preparation. The remainder of the kitchen team lined up the pinsas for cooking and cut them into bite-sized pieces for serving. They looked marvelous, and I was certain I'd soon be seeing raised eyebrows and hearing positive chatter as slices were passed around.

I wasn't disappointed. At tastings, noise is a good thing. Silence is deadly. That noise was building as more and more slices made their way to our guests. Several called either Anne or me over to explain the pinsa product and the process.

This was our curtain call. Heads were nodding. Yum! executives were smiling. And while it was no Super Bowl locker room celebration, we'd hit the nail on the head, and we knew it. This feel-good spirit lingered for a long while, even as the room started to clear. We'd accomplished our mission and experienced that rush of adrenaline we felt we had earned. Zefiro was a dead solid hit of the highest magnitude!

* * *

The next few days were a blur. I can't tell you what we did the moment the tasting was over. Did we go out and celebrate as a team? Did we spend the afternoon cleaning the kitchen and tending to follow-up details?

What I can tell you is that we did earn our pay, and the sense of relief and self-satisfaction was as strong as any period I'd ever experienced in my career. I'd made Christophe proud, forged new friendships, and cemented my reputation in the Yum! hallways as someone who delivered. There's no better feeling. It really is that simple.

Of all the dozens of small and large projects I've been a part of over the years, I've never felt that I'd done exactly what I was asked with the same level of precision I achieved with Zefiro. It is, to this day, one of my proudest accomplishments.

So why aren't I closing this chapter and heading out to my local Zefiro for a great pinsa romana? I can't answer that question. I can tell you that the world of franchising makes my head spin with all the legalities, financial machinations, and tensions. It's a confusing brew.

We built a beautiful space capsule of a concept. But getting it to Mars is a whole other mission. Zefiro sat on the launchpad for months. Eventually, the motor was shut down, the launch pad was disassembled, and the capsule was stowed away. Does it exist somewhere under a different name? Is it stored in a vault guarded by security cameras? I can't say. What I can say is that it's a remarkable story, and as Forrest Gump would declare, "That's all I have to say about that."

* * *

Postscript: Just months before publishing, I learned from Chef Ana Maria Rodriguez, a twenty-five-year veteran of Yum! Brands, that in the years following our Zefiro work, the company figured out how to reformulate the pinsa dough. They did it using sourdough culture and dry yeast, combining the mixture with the proprietary wheat flour used in all Pizza Huts globally. They dubbed the formulation "San Francisco style" and introduced it in more than twenty-eight countries globally, though not in the US.

Internationally, the response has been fantastic, with some countries now selling close to 70 percent of their pizzas with this new, lighter "San Francisco style" dough. She also shared that they were now experimenting with a new sandwich line using this special dough—something we had presented with our Zefiro work years ago. Ana's enthusiasm for the product allowed me to end this chapter as I'd always hoped.

Our work was successful after all. It's rewarding to put this in the win column. Maybe, one day, my friends and neighbors will have their chance to experience this remarkable "San Francisco style" pizza. Maybe.

13

Pizza in China?

A bullet train journey from Beijing to Shanghai takes a little over four hours, traveling at an average speed of one hundred and sixty miles per hour. The trip takes you from downtown Beijing to downtown Shanghai. Both the train station in Beijing and the one in Shanghai could be cities unto themselves. They are massive structures that challenge the imagination as to how these behemoths could possibly function.

Growing up in Little Falls, New Jersey, we had a train station at the top of Long Hill Road that my dad used to take to his office in New York City. That was always the train station I knew and my idea of how train stations worked. It had maybe five benches and a platform that was fifty feet long. There was a vending machine in the lobby and just two sets of tracks—one for arriving trains and the other for departures.

But departing from Beijing required a personal guide. He escorted me through layers of train station structure, down dozens of escalators, and past a Mall of America-like assortment of retail stores and eateries. We wove through a crowd whose density rivaled any of the busiest airports I'd ever been in.

My guide bypassed all the usual checkpoints and took me right to my seat on the bullet train in first-class into one of the most luxurious cabins I've ever seen. I speak zero Mandarin, and though my personal guide spoke a bit of English, our dash through the train station had us exchanging only hand signals. Once I was comfortably seated, we shook hands. And like a ghost, he was gone. I was ready to enjoy my first bullet train ride. Even as a veteran train traveler at the age of sixty, I was excited.

I'd been sent to China by Yum! Brands to look at their full-service Pizza Hut restaurants throughout the country. The principal executive was my old friend Micky Pant. The Yum! offices were in Shanghai. I'd never been to China, and while I was thrilled to work with Micky, I was aware of how starkly I would stand out among the Chinese due to my size and inability to speak the language.

My work would focus on Beijing and Shanghai, with excursions by car to smaller outposts like Hangzhou. Micky had assembled a team for me to work with, which included the smart and very perky Candy Chan, a long-time Yum! Brands executive. She had been educated in the US, and I relied on her to help me communicate with the rest of the team.

Yum! Brands had been one of the first Western companies to bring restaurants to mainland China, introducing Kentucky Fried Chicken to Beijing in 1987. Pizza Hut followed in the early 1990s, and by the time I arrived in Shanghai, there were over one thousand, seven hundred Pizza Huts in China. It was a massive company with a huge footprint and a loyal following. This would be a challenge for me in so many ways.

* * *

Traveling to China is very different from traveling to other countries. It takes a work visa, and there isn't the melting pot of peoples and nationalities you often see in other international destinations. Fortunately, there are a few nonstop flights from Dallas to both Shanghai and Beijing. But as with any first-time visit to a new place, I was anxious about what I was about to face.

It was a brutally long fifteen-hour flight. After arriving at Shanghai Pudong International Airport, Shanghai's massive airfield, I was fortunate to tag along with a seasoned American traveler who guided me to and through security and immigration. After the usual scrutiny of my paperwork, I made it to baggage claim and began to search for my contact whom I had hoped would be holding a placard with my name on it.

No such luck. There was an endless sea of faces in the welcoming area, and I towered over most of them. Having traveled in Asia before, I was prepared to be the oddball in the crowd. But I have never experienced anything like I what I experienced at Pudong. I was completely lost, completely incapable of communicating, and rapidly running low on adrenaline.

It was five o'clock in the evening in Shanghai. And having crossed the International Date Line, I was well into the wee hours of the morning according to my Central Standard Time biorhythm. I had no real concept of time, and the confusion at the airport made my head spin. Fortunately, I had the cell number of Candy, my contact. I called her.

No answer. I left a message and a text for her. I did my best to remain calm. Those of you who might chide me for not investing the time to learn basic Mandarin should know that while English has twenty-six characters representing vowels and consonants, Mandarin has well over three thousand. And despite my history of travel and intimate connection to several Asian restaurants in Dallas, immersing myself in Mandarin simply wasn't an option.

My phone rang. It was Candy apologizing for the confusion. My driver would be there in fifteen minutes, she said, as he'd run into unexpected traffic. I eventually found him waiting for me, and he bowed deeply, expressing a profuse apology. He spoke no English, but we communicated well enough for me to know I was in the right vehicle as he shuttled me out of the madness of Shanghai Pudong International.

What I didn't know was that we had an hour's drive into the city, all without any conversation. I did my best to just breathe and prepare for what was next. Shanghai is a massive city, with a nonstop skyline lined with residential apartment towers, high-rise office buildings, and bridges. The streets were filled with masses of humanity. It's ranked as the world's most populous city at twenty-five million people—almost three times the size of New York City.

Shanghai was overwhelming. And here I was squeezed into the back seat of a car with nothing to drink or munch on, no idea what the name of my driver was or where I was going. The faith you must have to keep these circumstances from triggering a severe anxiety attack is enormous. But my default justification for every job like this I took was that it was a way to pay the bills. And the bills never stopped.

Entering downtown Shanghai and firmly amid the last remnants of rush-hour traffic, I caught sight of a building with the familiar Marriott logo. We slowed and approached the entrance. I was relieved. The hotel staff was fluent in English, and I was directed to my small-but-comfortable corner room. I texted Micky to let him know I had checked in, and we arranged to meet in the lobby bar. We had a drink and some bites, but soon, Micky said he needed to call it a day. He gave me some quick instructions on the next day's agenda.

My mission for the remainder of the evening was to get a sense of the city, even if it was just to walk a few blocks to clear my head and stretch my legs. I always attempt to find a nice spa when I travel across the ocean. Chef Mark Miller, a seasoned international traveler, shared one of his travel secrets: Before you go to bed after a long flight, get a great local massage.

So I asked the front desk for recommendations on where to unwind and get a good massage. They highly recommended a few places, and the front desk staff offered to write my massage request on a piece of paper in Chinese to give to the spa attendant. With paper and directions in hand, I set out on my spa journey.

Walking into a street spa—not to be confused with a hotel or resort spa—can be intimidating, especially when you don't know the local language. I was approached by a diminutive older woman, and after giving her my instructions, I was led to a small room bathed in low light. It was thick with the scent of lemongrass. There was very soft music and the requisite slippers and robe.

It was impossible to get my size-thirteen feet into the slippers, and the robe wore more like a bath towel that had no chance of covering my body. The masseuse was a younger but also very tiny woman who seemed a bit surprised by my visage. She smiled, gave a subdued laugh, and motioned me to the table, offering to find me a larger towel.

Over the next hour, whatever tension I did have was completely exorcised from my body. I was reduced to a flaccid mass with no sense of time or space and no idea where my masseuse had gone. She was a ghost. I never saw her again. After a short rest, I dressed, made my way out of the room, and paid the bill. It was a wonderful start to my week in Shanghai.

The following morning was my formal introduction to the team at Yum! China in their offices just a stone's throw from my hotel. The offices were in a high-rise built over the Grand Gateway Mall, a shopping mecca filled with high-end luxury brands and one each of Yum!'s concepts in China: Kentucky Fried Chicken and Pizza Hut.

* * *

My consulting work required me to make a one-week trip each month to Pizza Hut locations in and around Beijing and Shanghai over the next seven months. I was to assist the team in their efforts to reposition Pizza Hut's full-service restaurants. I was offering my creative vision to upgrade and improve these operations. Key to this work was helping them become more competitive in a pizza market that was becoming more crowded with competitors.

For years, Pizza Hut was the dominant player in the Chinese market—a behemoth of a brand expanding throughout the country while generating one-third of Yum! Brands total revenues globally. Micky was called upon to direct this conceptual renovation, and because of our relationship, he had confidence I could be a strong and valuable part of the team. My previous work with Anne Fuller on the Zefiro concept and all the research we had done gave me a strong insight into Pizza Hut's challenges. Or so I thought....

Our first stop was to the nearby test kitchen of Yum! China Pizza Hut. We were escorted to a tasting room set for twenty-five people. I've attended and conducted dozens of tastings over my career, but never had I been to a tasting event with this many people. We would be sampling more than twenty items ranging from appetizers to pizzas to desserts.

With no clue as to the protocols and knowing I wouldn't understand a word anyone was saying, I relied on Candy to interpret and inform me of anything I needed to do. I soon realized I wasn't expected to do anything other than observe the process. I'm not sure how long the tasting lasted, but I still have photos of the food.

Servings included a pizza topped with cotton candy; a pizza with fried chicken embedded in the exterior crust; and one pie with what looked like tater tots sprinkled over the entire pizza. It was open season on whatever the teams could dream up. I looked around the room, and the group seemed genuinely engaged with each strange—to my eyes—variation.

It was all a bit surreal for me but not at all out of the ordinary to the people in the room. There was consensus agreement as to the appeal of each dish and even a few rounds of applause. I did my best to smile and not draw attention to myself.

I had time with Candy and Micky to discuss what we'd just been through following a thorough tour of the facility. I found out that this type of exercise wasn't all that uncommon for the team. For years prior to Micky's leadership position, Pizza Hut in China was directed by a dynamic and visionary leader Sam Su, a Chinese national who was enamored with Western influences, flavors, and trends. He'd earned an MBA at the prestigious Wharton School of Business at the University of Pennsylvania.

Pizza Huts in China had menus that would strike most Westerners as odd. But Pizza Hut China was not so much a pizza restaurant specializing in pie platforms as it was a TGI Friday's with menus that dramatically expanded traditional pizza boundaries. Pizza Huts in China offered items like escargot, smoked salmon, sous vide octopus, beef curry rice, and grilled sirloin steak. An American tourist in China might recognize the familiar logo, but should that tourist step inside, the similarities abruptly dissipated.

I counted eighty single SKU items—ingredients that were used for only one dish with no cross-utilization over the menu. For example, purchasing frozen snails for escargot and an exclusive, out-sourced garlic butter sauce would be unheard of in a typical restaurant—especially if that menu option was not a big seller. These are what I call albatross options: they weigh down the operation with purchasing, storage, and production requirements and are nearly impossible to prepare and serve with any consistent quality.

In the US, these options are quickly culled from the menu once that pattern is recognized. In China, if Sam Su said he wanted the dish on the menu, it stayed on the menu regardless of cost. I'd also learn that the previous leadership team loved traveling to France and other international markets.

This menu started to make more sense as I received more feedback. It took me a while to get used to the almost incomprehensible menu diversity, but I had to understand the "whys" of it before recommending any changes. It was delicate work. Over those seven months of work, my goal was to absorb and assess as much information as possible. I would then prepare a final report with recommendations on how the one thousand, seven hundred-plus Pizza Hut China locations could be more competitive and align with the changing demographics of the various Chinese markets.

No small task. The diversity was eye-popping. Some drink menus offered up to forty different teas, soft drinks, smoothies, and frozen cocktails, along with bottles of wine and champagne. Dessert menus ranged from matcha ice cream cake to waffles with ice cream and fruit. Locations ranged from super slick nightclub-like restaurants in Shanghai with bouncers minding the doors to smaller mall locations in industrial cities not much different than a Pizza Hut in Wichita, Kansas. (Wichita is where brothers Dan and Frank Carney started Pizza Hut in 1958.)

The variety of these locations and the monstrous gaps in sophistication among their customers were mind-boggling. In one location in a very rural area between Hangzhou and Shanghai, I was stopped ten times by locals who wanted to take a picture with me because (my guess) they had never seen an American Caucasian ever—especially one my size. The dumbfounding reality is that each Pizza Hut had more than four hundred and eighty items on their order guides—a number that would make any chef I know choke on his morning donut!

But this was the challenge. Pizza Hut China had been expanding the main brand and its menus for years, even as they were technically part of the Yum! Brands corporate portfolio. Yet it acted more like a franchisee than a company-operated brand.

* * *

This was the challenge Micky faced. How do you pull back the reins after years of expanding locations and menus and perform as a consistent brand? For decades, there had been no credible competitors to the Pizza Hut brand in China. There were no competitive guardrails to keep a free-for-all at bay. Without that pressure, companies tend to drift, knowing that mistakes won't harm them significantly.

If I, as a guest, selected a menu option that sounded interesting but ultimately wasn't worth the trouble, cracks emerge over time in the conceptual framework. At first, they seem harmless, but as those disappointing experiences accumulate, two significant consequences surface.

First, revenues dip as guests seek alternatives. Second, competitors aggressively seize on the opportunities embedded in those declines. As I toured around these locations, it was obvious to me that Micky understood this too. Ignoring these potential pitfalls could be a serious problem. Pizza Hut China was the six hundred-pound gorilla and, while still firmly in charge, was starting to age. In the restaurant industry, revenue is king, and when it starts to flatten or growth slackens, alarms start buzzing.

Micky was aware of the work I had done with Yum! on Zefiro concept development. This concept was the direction I felt the industry was moving. This kind of innovation was at the heart of my work. During my short stint in China, I was keenly aware that the crew I was working with was not a creative team and that Pizza Hut China did not have the flexible internal culture necessary to spark innovation.

It wasn't my role to consider all the politics involved, and they were considerable. Rather, my focus was to effectively position Yum! China to capitalize on their market leading status—give them a game plan and let them execute. Nowhere was that more evident than in ingredient acquisition. The director of purchasing for this massive restaurant system was painfully aware of the challenges and pitfalls associated with procuring ingredients from overseas. Pizza Hut China imported all its cheeses from the US. You can imagine the stunning breadth of procurement work required for one thousand, seven hundred-plus restaurants!

Add to that the challenges of sourcing ingredients for odd pizza menu items like osso bucco, calamari, snails, and a variety of steaks. It was obvious the director of purchasing considered success in his position to be all but impossible. Why I ever imagined I could help break this logjam is beyond me. But I was being paid generously to offer my opinions and ideas, and that's exactly what I did.

Once again, I brought Chef Mark Miller on board. His familiarity with all things China and his uncanny ability to understand cultures and palates were crucial to creating a road map out of this dilemma. My instincts told me to keep it simple. I wanted to challenge the Pizza Hut China team to put aside their broad, eclectic menu and focus on core items such as pizza, pasta, and Italian entrees. This would reduce the time it took guests to plow through the menu—currently at twenty-plus pages, book-bound with a hardcover!

I wanted them to think of Pizza Hut as a modular brand that could easily adapt to different demographics and compete locally, paring down menus in locations where sophistication levels were far different from the bustling urban enclaves of Shanghai and Beijing. Sounds easy enough, right?

Over the course of the next few months, I researched and reviewed—both in China and in Dallas—articles and key historical data. Chef Miller and I spent time eating and discussing Italian cuisine in addition to debating which peripheral menu items worked. Kitchens in the Pizza Huts throughout China were burdened and busting at the seams. So, our work had to include kitchen capacities and layouts.

It was a monumental task. But it led to a presentation that, to this day, was one of the most unusual of my life. I had composed a series of modular menus with a small team based in Dallas. Another member of my team, Amber Brown, assisted with developing modern menu presentations and graphics. Mark and I collaborated on menu choices and descriptions. Our pizza modules allowed for a variety of crust options, fresh salads, and versions of Italian classics like lasagna and traditional pasta dishes.

A friend, Chinese by birth but raised in the US by Chinese parents who didn't speak English, translated all my modules into Mandarin. I thought of everything. Armed with a small suitcase packed with a dozen completed menu modules in both languages, I headed off to Shanghai for a three-day stint to make my presentation. I was excited and anxious to showcase this new vision to a team of twenty people the day after I arrived.

I'd long since changed my hotel to a more upscale Marriott at Tomorrow Square and settled into a night of walking The Bund—a waterfront area and a protected historical district in central Shanghai—to gain energy for my presentation. My comfort level had increased, and now, I felt like a regular in Shanghai, walking the streets without anxiety. I discovered small noodle shops, dumpling stalls, and even a Jewish deli called Tock's with ties to Montreal. I'd sampled several great street-side massage spas and no longer needed my original written instructions.

I'd met most of the people I'd be performing for and had rehearsed my presentation to the point I felt I could field all questions. The menu module approach was innovative yet not difficult to explain. The concept was based on matching modules to the demographics of the various locations. I focused on a small, rural Pizza Hut location outside Hangzhou that didn't need the same menu as an ultra-urban location in central Shanghai.

But they did need the same core menu of pizza styles, salads, drinks, desserts, and appetizers. Larger urban locations could add modules that offered more international choices but without esoteric options like octopus, snails, and salmon tartare. In all, I had nine different modules and had managed to trim the overall menu by 30 percent. It was about a three-hour presentation as I played it out, and at nine o'clock in the morning the next day, I was ready to roll.

I'm not sure how many of those people understood a word I said, but I had their rapt attention throughout. Not a single person exited to use the restroom or get a refreshment. I gathered momentum as I presented the entire module strategy and spoke without interruptions or questions.

When I was finished, I looked over to Candy Chan and Peter Kao, the two most senior team members and my direct contacts. I asked the people in the room if they had any questions. No response. It was as uncomfortable a period as any I'd ever experienced. Candy started to clap. Then the entire room erupted into applause—albeit just slightly louder than a golf clap.

It lasted maybe ten seconds. Peter stood up to leave the room, and everyone but Candy followed in his footsteps. She remained behind and thanked me for my time and the thoroughness of the presentation, then quickly left. I stood alone and suddenly realized how abruptly my seven months of work had ended. I collected my things and waited a few minutes to see if anyone would return to discuss the rest of my visit.

Crickets. No high fives, no "Let's grab lunch and a drink" or "Call me later, and we'll have dinner." My work was finished, and that was that. I spent two more days in Shanghai and never heard another word from anyone at Pizza Hut China.

I'm not overly sensitive to negative or ambivalent responses to my work, at least not more so than anyone else I know. But I was startled by the silence. I left the conference room alone and departed the offices without any further contact. I spent those two extra days as a tourist—a perfectly enjoyable ending to the "out of the box" work I'd done and the confusing response.

On that last day, the air was filled with snow flurries as I pulled up to my American Airlines terminal at Pudong International. I pondered how every snowflake is unique, each with an entirely different structure. I smiled, knowing I would probably never see a job like this again. Unique. An understatement.

* * *

Footnote: A few years later, in 2018, Yum! China converted the China operation into one large franchisee with Chinese ownership, quickly making Pizza Hut China the largest franchisee in the Yum! Brands portfolio. In October of that same year, *Restaurant Business Magazine* published "Take a Look at this New Pizza Hut in China," an article on the unveiling of the Pizza Hut prototype that had just opened in a fashionable part of Shanghai.

The article noted that same-store sales had been weak, with a 4 percent decline in the second quarter ending June 30. Yum! China said the new restaurant was designed to deliver a modern and fashionable dining experience, one likened to a private meal in a chef's home. I smiled when I read that this new store would enable the company to test new menu innovations before introducing them nationally.

"The chain plans to open more such stores in different parts of the country to gain deeper insight into customers ever-changing preferences," the article concluded.

As of this writing, there are now more than two thousand, three hundred Pizza Huts in China. That's six hundred more than the number of Applebee's locations across the globe and 50 percent more than the number of Chili's Grill and Bars in the United States. Between Yum! Brands Kentucky Fried Chicken and Pizza Hut in China concepts, there are now more than ten thousand of these restaurants in one thousand, five hundred cities throughout this Asian nation.

That's four times the number of McDonald's in China, and these restaurants generate $8 billion annually, making it, according to the franchisee Primavera website, the number 383rd company in the Fortune 500 as of 2020. I admire how groups like Yum! China use whatever tools they have at their disposal to build a powerhouse company.

Funnily enough, I still don't know if they used my modules, or if you can still order escargot from the menu in the new Pizza Hut Shanghai prototype. But I still have a set of Pizza Hut menu modules perfectly wrapped in my "Yum! China" storage box.

To this day, I consider those modules to be some of the best consulting work I've ever done. In some small way, I believe I may have had an influence on the dining trends in the most populated country in the world—almost seven thousand, five hundred miles from New Jersey and a long way from where I enjoyed the first slice of pizza I'd ever had.

14

Food Halls. The Next Culinary Rage

It must have been well past nine o'clock at night when I landed at Honolulu's Daniel K. Inouye International Airport. The seven-hour flight from Dallas travels through five time zones and involves a good deal of disorientation, especially in the darkness of an airliner cabin. Fortunately, I had negotiated with my newest client to travel first-class—a practical consideration given my size and self-diagnosed claustrophobia. I had been to Honolulu just once before and didn't know my way around.

I got a cab, took the twenty-minute trip to downtown Waikiki, checked into the Hilton Garden Inn, and put my things away in my very small room. It was early enough that my new client's food hall would still be open. I was anxious to experience it for the first time. It was an orientation trip of sorts, where I would observe, meet some of the managers, and get a feel for the environs.

I walked across Kuhio Street and found myself in the middle of The Street—A Michael Mina Food Hall. My client? One of the most accomplished, decorated, and respected chefs in the United States: Michael Mina. This was his first venture in the food hall business, and I finally had gotten a chance to work with him and his team. It was a stunning space, even late at night. I smiled and set out to understand what he had created here in paradise.

* * *

I'd gotten into the food hall business mostly by chance. My work with Trinity Groves, Velvet Taco, and Banh Shop thrust me back into the limelight. I was getting calls from people I knew in the business. One of those was Darin Botelho, the retail vice president with the Gables Management Company, a Dallas-based luxury apartment developer. High-end apartments often come with upscale amenities, such as full-service restaurants to help lure both tenants and well-known chefs.

Unfortunately, it wasn't uncommon to find these restaurants empty. In Dallas, unlike New York or Chicago, restaurant-goers have never been keen on making these apartment-attached restaurants a regular stop. The reasons are numerous, but primarily, it boils down to parking. The reason for Darin's call? He wanted my opinion on what to do with a seven-thousand-square-foot space on a highly visible corner in Uptown Dallas. It was home to at least three failed restaurants over the past ten years.

Ironically, it was the same building I had considered to open Bengal Coast five years earlier. But management feared the odors from an Indian restaurant might befoul their luxury dwellings. Go figure. Anyway, Darin wanted to brainstorm ideas with me about what to do with the space that wasn't just another restaurant. We met and discussed some of the trends in the business, with a focus on how to convert the existing space to something trendier and more versatile.

Darin and I concluded that the space was better suited for a food hall, an emerging trend getting a lot of attention in the industry. The space had a full restaurant kitchen in place, vent hoods, bathrooms, and a gorgeous patio. All the bones were in place. Food halls are essentially upscale food courts, not unlike those you see in shopping malls or airport terminals. In highly dense urban areas like New York City, Philadelphia, or Boston, food halls had been around for many years. They thrive because they offer chef-driven choices for people on the run any time of day.

Every little town or city center along the East Coast where I grew up was sprinkled with small ethnic restaurants. Even in our little downtown of Little Falls, New Jersey, you could get pizza, gyros, Chinese food, Jewish deli fare, or a great burger at the luncheonette, all on the main drag. Food halls pack all of those choices together in one building.

In Atlanta, two of the most frequently mentioned food halls were Ponce City Market, an old Sears store conversion, and Krog Street Market, a smaller one-level space with a neighborhood feel. Darin was smitten with both. His company, Gables Residential, had several apartment complexes in Atlanta, and when his business took him there, he always gravitated toward those markets. Dallas didn't have anything like that, and he saw a great opportunity to experiment.

He invited me to join him on an upcoming trip to Atlanta to review those spaces, among others, to get fresh ideas for Gables Residential's abandoned seven-thousand-square-foot space. I had spent time in Atlanta on several occasions and was familiar with the food hall scene there. And I'm always a sucker for an all-expenses-paid trip to great places to eat and drink.

We ate and drank, passing a full day in Krog's Market, Ponce City Market, and Westside Provisions. With each stop, we tried to envision how to bring some of that magic to Dallas. There were great baked goods, cool bars, craft beer, hot chicken—a trend just getting started—and dozens of other specialty foods. It was a feast for the eyes and the belly.

By the end of the trip, Darin was convinced that his space in Dallas needed to be a small-yet-energetic urban food hall. Yet, at sixty, I was not fond of running such an operation: Too many headaches plus too many unknowns with no help running it equals potential disaster. But Darin's formula was attractive. Gables would provide me with twenty-five dollars per square foot to build out the space any way I saw fit, provide a no-guaranty lease if it doesn't work; access to their residential tenants, and they'd maintain the property without any fees. How could I lose?

Beyond the $175,000 Gables was willing to commit to convert the space, I needed another $100,000 for design, which I did not have. I needed a partner or some other financing. I tapped Brad Woy, whose company CapRock Services lent funds in exchange for managing direct day's receipts deposits through Automated Clearing House network payments made nightly. CapRock was just one of many companies doing this, supporting their loans through transaction fees in all credit card processing as a way of whittling down the debt over the term of the agreement.

I knew Brad from a pitch he had made for these services at Trinity Groves. Phil Romano didn't need that kind of service, but Brad and I had kept in touch. Though we'd owe him the $100,000 plus interest, we calculated that we could pay off that loan within two years of operation. I also started looking for a partner who could operate the space and organize seven chefs and restaurateurs to share our food hall vision.

The space had a central kitchen, a wood-burning pizza oven, and two bars and needed few modifications. We designed a comfortable outdoor space, added a conference room to rent out for presentations with access to our food and beverage services, and brought in great signage. We called it Urban Market, "UUM" for short.

* * *

Around this time, my attorney Ira Tobolowsky asked me if I would go to lunch with his three sons: Jonathan, the oldest; Michael; and Zach, the youngest. He'd been trying to get his sons started on careers and suggested that I might have a talk with them over lunch. Ira was like a father to me. His kind demeanor, intelligence, and deep insight guided me through good times—Canyon Café and Pei Wei—and bad—Bengal Coast and my divorce from Jana.

I thought of Ira as family. So when he asked me to meet with his sons, I was both honored and willing. His son Michael wanted to be an attorney like Ira, and Zach was still at Yale, undecided on where to focus. He was probably along just for the ride at this lunch. It became clear to me that Ira's interest was Jonathan. He was, at that time, looking for his next opportunity.

Because of my history with Ira and his respect for what I'd accomplished in the restaurant biz, he was hoping I could get something going with Jonathan. He and I hit it off well, the Gables project was off to a good start, and I still needed someone to run it. So Jonathan and I agreed to talk the following day and walk the space together.

Ira was ecstatic. He called me that afternoon to thank me and agreed to do the legal work for free. I told Jonathan he could own 75 percent of our partnership, but he'd have to do 90 percent of the work. I figured I could keep consulting and that UUM could sustain Jonathan's salary needs easily, and if there was any leftover, I might wet my beak too. Jonathan and I set up a new entity to manage the project, and Ira did all the paperwork.

I secured a $100,000 loan from CapRock at a hefty interest rate with an agreement to have them manage our credit card receipts. They'd have automatic withdrawal privileges from our bank account to ensure they were getting their loan payments, an agreement that came with a personal guarantee. I approached Jonathan about us splitting the guarantee fifty-fifty. But his counsel advised him not to sign any guarantees. It was my decision to make.

Being the perpetually "glass half full" person I am, I signed the guarantee. I figured that in the worst-case scenario, repaying $100,000 over a few years was doable. Food halls are generally not operated by one restaurateur or chef. They are a compilation of small spaces with many chefs and restaurateurs chipping in. It's much like a co-op, with all owners relying on each other to make things work.

The collective strength of these ideas is the core attraction. With the Gables money and my $100,000, I felt I could make this work. I had a strong network of operators, and the only similar offering in Dallas was our Farmer's Market, which had food stalls open primarily on weekends. With three thousand Uptown area residents and a location at a busy and recognizable intersection, we couldn't fail, right?

Within a few weeks, I was able to attract a who's who list of potential candidates to fill our small food hall. We landed Taco and Dunia Borga from the established La Duni bakery; famed chef Tiffany Derry; Buda Juice; pizzeria chain Fireside Pies; acclaimed chef Gilbert Garza; and a couple of hip young bar operators: the Beardon brothers. I deemed it critical to attract locally known names with strong reputations to help promote this new idea.

It's not enough to have a broad range of foods. It takes word of mouth spread by those knowing the names and reputations of the chefs and restaurateurs involved. This was that kind of collection, and I was confident each of these professionals would bring their own following to the project. In addition, we were offering pizzas, breakfast pastries, coffee specialties, tacos, burgers, and two great bars along with a stunning patio.

The formula was well designed, and Darin and Gables management were optimistic we'd done our homework. It was cool. It was different. It was fun. But it was the wrong place at the wrong time.

That pretty much sums up UUM. There was nothing technically wrong with the concept. So why was the timing wrong? The Dallas consumer at that time did not understand or have a point of reference for food halls. Even today, with all the hubbub about food halls over the past six years, there's only one strong example operating in the entire Dallas-Ft. Worth market: Legacy Food Hall in Plano just north of Dallas.

Many others have opened and closed. Just a few have seen moderate success. And most of those are kept alive by deep-pocketed landlords or developers who need food as an amenity to their offices or high-rise complexes. Dallas doesn't have the urban culture to shore up concepts like this.

I was not going to UUM daily, though I did stop by often, convinced success would arrive if given time. Jonathan was in charge, but his lack of restaurant experience was showing, and the team of licensees started to call and email me complaining about his management style. Little by little, these cracks in the management cement turned into gaping potholes.

Jonathan and I met a few times at UUM to discuss his operations approach, and it got confrontational. He would be combative over the simplest of decisions. And given my financial obligation to CapRock, I grew impatient with his attitude. I'd chide him for coming to UUM in shorts, a tee shirt, sandals, and a ballcap put on backward. The operators took note of his lack of experience managing restaurants and chefs.

Jonathan would snap back that he had majority ownership, forgetting that I had gifted him that percentage without requiring him to take on any financial risk. But rather than pull the plug, I decided to stay away and hope for the best. We were still relatively new, and my hope was that as word got out, our business would grow, and I could afford to get more involved. But that approach didn't sit well with the licensees.

They were rebelling against Jonathan's style, and some threatened to walk out due to his lack of leadership. It was a difficult time. So I offered to find someone to buy Jonathan out and run it. Even though Jonathan had no financial risk, he felt his ownership had value because of the time he had put in without drawing a salary. It was a standoff. I didn't owe him a plug nickel, but we'd have no business to manage if things didn't improve. I had to get him out somehow.

My solution was ex-Pei Wei General Manager Scott Koller. He was smart, experienced, and immediately likable. I knew he was looking for a new gig and had a few dollars to invest in a new opportunity. I invited him to come by and see UUM, and we spent a few hours drafting a plan to buy Jonathan out. Scott was sold on the idea of coming in as a partner, and while I tried my best to present the positives and negatives of his decision, it was ultimately his call.

Negotiating with Jonathan was extremely difficult. I had to invite several people, including his brother Michael, to hammer out a workable solution. It was impossible to precisely determine his ownership value because we weren't making any money. Plus, Scott established limits on what he was willing to invest. After some exhaustive back-and-forth, they came to an agreement to buy out Jonathan's 75 percent stake. He hadn't generated any value through his ownership position, but it was the only way to move forward.

Scott did a remarkable job gaining the trust of our food operators, and it took little time for him to develop a positive culture. He would work in the kitchen, mop floors, and greet and help new guests make decisions. UUM became fun again, and I spent more time there, trying to figure out how to boost sales. But even with this new and improved vibe and The Gables support, UUM never turned the corner into profitability. After six months, we decided to pull the plug.

* * *

It was another fine mess I'd gotten myself into. Besides the failure of another venture, I had that outstanding CapRock loan to pay off and not enough income to meet the debt service. I worked out a payment plan and eventually was able to close that loan. Scott took a financial loss as well, but he was not bitter about the results of his choice. He understood the risk.

To exacerbate the situation, Ira had passed away suddenly during this maelstrom. He never had a chance to intercede. I would never see Jonathan again. It was just a bad chapter and experience to draw from. Again. Yet I was able to build on the positives from this episode. Russell Friend, another colleague from my Pei Wei years, came to Dallas from California to see my work in the food hall sector. Russell left Pei Wei not long after I did and landed a lucrative position as the chief development officer for The Habit Burger Grill based in Irvine, California.

Habit had been a darling of the burger world. After some strong years with Habit, which included an immensely successful IPO, Russell decided to leave with a nice pile of cash. He fell in love with UUM and saw the same opportunities for growth that I had seen. He felt we could launch a consulting company to help larger retail space owners and developers figure out how to convert large building spaces into food halls.

Facing fierce competition in the online arena, brick-and-mortar retailers were abandoning mall and anchor shopping center spaces at a fast clip. There was opportunity. Russell stepped forward with funding, and we got the ball rolling with VisionHall, the name we gave our consulting company. We were able to secure some consulting gigs, and while not as broadly successful as we'd hoped, we did get our names in front of all the big mall operators: Macerich, Westfield, Taubman, Brixmor Property Group, and many others.

Over a two-year period, we cobbled together enough work to support both of us and saw opportunities to expand our work. But not everyone wanted to hire VisionHall. Several people wanted to hire us individually. That made sense since Russell and I had vastly different backgrounds: real estate development (his) and concept development (mine). It was a good balance, but clients didn't always want or need both.

One of those clients was Chef Michael Mina, who had just opened a food hall in Honolulu. I had met Chef Mina through Rick Federico, who had crossed paths with him through the Bourbon Steak restaurant Chef Mina had launched at the Scottsdale Princess Hotel. At Bourbon Steak's grand opening, Chef Mina shared with Rick his passion for creating smaller, more casual restaurant spaces.

Rick had always respected my work in this genre and passed my name to him. Soon after, Chef Mina invited me to come to see him at his eponymous restaurant, Michael Mina, in San Francisco. I was treated to a seven-course meal that Chef Mina himself prepared and was dazzled from start to finish. We hit it off well and followed up with more meetings. But the timing just wasn't right. (This was years before his food hall in Honolulu.)

Mina and his team were in high demand for fine dining development around the country, and they were being wooed by some of the top developers in Vegas, Boston, Hawaii, Seattle, and California. I wasn't a fit for these projects, but the relationship was forged. And eventually, an opportunity emerged.

Billy Taubman of the Taubman Group, the largest mall operator in the US, had approached Michael with an offer. He wanted him to open a high-end Michael Mina restaurant called Stripsteak on the Lanai level in the renovated International Marketplace shopping mall in downtown Waikiki. Along with Stripsteak, Billy was enamored with the growing presence of food halls in large failed or abandoned mall spaces.

Billy and Michael worked a deal to open The Street—A Michael Mina Food Hall, on the ground level of Taubman's International Market Place mall. It featured seven of Chef Mina's conceptual ideas for fast-casual cuisine: pizza, ramen, burgers, bars, BBQ, and poke. There was an Egyptian food stall called Little Lafa, a passion project for Chef Mina, given his Egyptian heritage.

The anticipation was enormous. Russell and I caught wind of it when we were doing our VisionHall work. I didn't think much of it because Honolulu was way outside of our operational zone. Then, one day, my cell phone rang.

"Mark, it's Michael Mina," he said. "How are you, and when can you come to Honolulu?"

Challenge: How do I tell Russell that the work I could potentially be doing for Chef Mina was a consulting project for me, not VisionHall? We had a productive conversation about the work, and while it wasn't an easy choice, we agreed that my work would ultimately help VisionHall if I had Michael Mina on our client list. This Mina project at least allowed me to have an income outside of VisionHall and gave me freedom of movement.

On my initial trip to Honolulu after that long flight from Dallas, I walked into the vast Michael Mina food hall for the first time. I quickly assessed what looked right and what didn't. Based on the light fixtures, furniture, and equipment, it was obvious that this was an expensive project. I immediately sensed I could offer constructive direction. I went back to my room at the Hilton Garden Inn and slept well, anticipating the next day when I would meet Michael's team on site.

I arrived at nine o'clock in the morning and was immediately greeted by a smiling face and a warm handshake.

"You must be Mark," came a friendly voice and genuine smile.

"Yes, and you must be Jonathan," I replied.

Chef Mina had given me some background on Jonathan Lee, the general manager of the food hall. He was a deeply experienced restaurant guy with a strong culinary background. I immediately liked him, not just because of the warmth he emanated but also because it was nine in the morning, and he was already hard at work. I was off to a good start, and, hopefully, this was a Jonathan with whom I could successfully interact.

Honolulu gets about six million international visitors each year, and most of them end up in Waikiki, the main beach and heart of the hotel district, with a view of the famous Diamond Head Park. I've often compared Kalakaua Avenue, the main drag in Waikiki, to the strip in Las Vegas but with a beach at one end. It's a constant source of traffic, sunshine, and activity. The Michael Mina food hall was connected to Kalakaua by a breezeway in the three-story International Marketplace shopping mall.

Each week, about one hundred and fifteen thousand new visitors land in Waikiki, spending $157 million or $1,400 a person per week. In 2019, the total visitor spend on the island of O'ahu—the third-largest of the Hawaiian Islands and home to Honolulu and more than two-thirds of Hawaii's population—was $8.2 billion. Every retail business in Waikiki is casting its net to snag a chunk of that spend.

Original projections for the Mina food hall estimated the space would generate $15 million annually. But in the first six months of business, they were falling far short of those numbers. I was standing in front of Jonathan Lee, trying to figure out why revenues weren't where everyone—including Billy Taubman—thought they should be. Tricky.

Chef Mina is arguably one of the top five chefs in the United States, and the food being served at the food hall was both delicious and inventive. But there was something missing. There must be a certain vibe about a place, a soulfulness that you realize immediately if only subliminally.

Great food isn't eaten in a vacuum. The music, seating, lighting, design, pricing, and hospitality all play a role. This food hall was culinarily focused to the exclusion of these other critical elements. Plus, the cuisine was too aspirational; not as approachable as it needed to be. The offerings were too esoteric for a tourist market looking for something fast and affordable in the sea of high-priced restaurants washing over Waikiki. Chef Mina's team had aimed too high.

Jonathan and I spent a few days together, reviewing all the different foods, and overate. I'd traveled about four thousand miles to get to paradise but never even dipped my feet into the warm Pacific Ocean waters. I ended up inking a long-term agreement with the Mina Group that had me traveling to Honolulu for one week each month over the course of a year.

My focus was on creating more accessible food spaces and reprogramming the dining areas in terms of seating, signage, and offerings. Collectively, we created new and exciting replacement food venues. These included a surf-and-turf concept serving sliced-to-order smoked prime rib French dips and a mouthwatering King crab roll. There was a hot fried chicken space called Uncle Harry's Hot Chicken. Great fried chicken satisfied the general strategy of offering accessible foods with the exquisite touch of Michael Mina.

We'd also taken a small kiosk space called Indie Girl and expanded it to provide healthier options to balance the offerings of The Street. In short, we now had something for just about everyone, created by some of the best culinary minds in the business at affordable prices. There was exhibition-cooking, aromas, and culinary action everywhere. The new changes created a buzz in The Street, and the staff responded with what Hawaiians call ho'okipa, a level of hospitality that is endemic to the islands.

Unfortunately, during the process, Jonathan Lee was forced to exit for personal reasons. He had maintained a homestead on the island of Maui, the closest island to O'ahu but still a short flight away. He had been commuting between islands, often not going home for weeks and staying in a small apartment near The Street. His family hadn't adjusted well, and he needed to invest more time with them and not the food hall.

We scrambled to find a replacement. The Mina Group turned to me for suggestions. I had a crazy idea. After my own food hall closed, I knew Scott Koller hadn't yet landed another position. I called him from Honolulu and suggested he temporarily relocate to Hawaii from Dallas for a salary that would curl his hair. Scott agreed and inked a short, three-month agreement to patch the hole we had in our team until we recruited a permanent replacement. It was as good a win-win as we could have hoped for on short notice.

Scott had never been to Hawaii. He would leave his home and family for a small furnished apartment to work seven days a week. With Scott's input and leadership, the food hall started to come together. We added Mi Almita, a Mexican restaurant on one of our more visible corners, after teaming up with Michelin-starred chef Hugo Ortega from Houston. After additional fine-tuning, it was up to the operating team to make the best of the changes.

Sales increased, and expenses decreased—the double-barreled foundation of all restaurant success. As my year ended, Scott's temporary status was renewed for another three months. Chef Mina sweetened the deal further by providing Scott, his wife, and his three young daughters his home on the north shore of O'ahu for a full week's vacation.

For me? I enjoyed the benefits of spending a week in Hawaii each month for a year with all expenses paid and dining on some of the best cuisine I'd ever put in my mouth. I'd also potentially cemented a long-term relationship with Michael Mina. It was more than I could have hoped for on that first late-night trip to Honolulu.

I departed the city for the last time after fourteen months of focused work. I'm embarrassed to say that I never did dip my feet into the Pacific Ocean or enjoy the gorgeous beaches of O'ahu. But I have no regrets, as I had the opportunity to work side-by-side with the most culinarily creative team I could have ever imagined. My memories of Hawaii, the great staff that welcomed me so warmly, and a year in paradise made it all worthwhile.

* * *

Back on the mainland, my food hall work with Russell Friend had dried up, and he and I had lost the steam needed to continue operating VisionHall. Russell was looking for a more permanent position, and with my hiatus in Hawaii lasting so long, we'd lost any momentum we'd had. We both moved on to other work. A huge chunk of that work would come my way via my old friends from Velvet Taco and Front Burner Restaurants.

The Front Burner Restaurants team had just signed a mega deal to develop a fifty-thousand-square-foot food hall in Plano, Texas— about twenty miles due north of where I had opened and closed my own food hall in Uptown. My relationship with Front Burner principles Randy Dewitt and Jack Gibbons was intact. I was able to ink a consulting gig helping them open Legacy Hall, their own vision for a super-sized, three-level food hall with an on-site brewery. It was a massive project.

The vision for Legacy Hall dwarfed anything I'd worked on before. But the essential elements remained the same: great food stalls, a lively bar, design-forward environs, and a strong urban sense of place. Legacy Hall also had a huge outdoor beer garden, with a monstrous stage for concerts and live events. Not a dime was spared. Typical Texas thinking.

Legacy Hall even afforded me the opportunity to create my own food stall with Pardeep Sharma, founder of India Palace and the godfather of Indian cuisine in Dallas. He was never a fan of my more modern approach at Bengal Coast, but we were friends nonetheless as he'd always respected my entrepreneurialism and dedication to my craft. He agreed to bankroll one of the sixteen licenses at Legacy Hall, and together, we created a great little stall we called Blist'r, serving Indian naan wraps and bowls.

Legacy Hall opened like wildfire. Lines formed on the sidewalk outside of the food hall for weeks. People had to wait just to get in, as publicity was enormous, stoking curiosity beyond the capacity of the space. I'd never experienced anything like that in my travels to other food halls, regardless of how busy they were. Between the collection of great restaurateurs and chefs, the opening few months were beyond expectations.

Word of Legacy Hall's successful opening spread nationally, and people arrived from all across the country to see what all the fuss was about. Revenues exceeded expectations, and Legacy Hall was out to prove that food halls didn't have to be deep urban to succeed. Everyone seemed very pleased and proud to be part of such a gigantic success story—that is, until the pandemic of 2020 reared its ugly head.

Food halls thrive on crowds and wall-to-wall activity in tight spaces. Standing in long lines waiting to order food is a given. Even with large outdoor seating areas, the pandemic brought mass-gathering social spaces like Legacy Hall to a screeching halt. One week, it was as crowded as a U2 concert or a Super Bowl extravaganza. The next week, crickets.

That revenue faucet was shut tight, and not a drop of cash dribbled out of that spigot for months—even a full year. I had moved on from my work at Legacy Hall when the pandemic hit, but my ties to the people remained intact. I continued working in the industry, so my interest in the effects of the pandemic on food halls never waned.

Blist'r is one of the four remaining original food stalls still open at Legacy Hall, and I'm proud of what Pardeep and I created. The truth is that I am not sure Legacy Hall will ever recover the magic it once had. There may be some recovery. But my guess is that, particularly in Dallas, we have seen the best of the food hall years.

Cities like New York, Chicago, Atlanta, and Seattle will continue to be the beating heart of this urban lifestyle trend. But the rush to fill vacated big-box spaces with large food halls and bars might have seen their day. In the restaurant industry, so prone to new trends, it's difficult to predict how these large spaces will evolve. People are still wearing masks as I write this, and not everyone will again be comfortable in the tight confines that busy food halls entail.

I doubt the trend is over, but it's not the shiny new toy it once was. For me, it was time to get back into the fast-casual business: the bread and butter of my career success. Time to move on to the Japanese powerhouse concept called Marugame Udon. The irony of this next move makes me smile as I write this. Why? That's a good story too.

15

Udon Succumbs to COVID-19

Back in 1964, when I was eleven years old, our family spent summers on Long Beach Island, New Jersey, a gorgeous eighteen-mile stretch of beach just north of Atlantic City. I'm the youngest of five siblings, all clustered together over a span of seven years. Along with my three older brothers, who were thirteen, fifteen, and seventeen at the time, I'd join my father, my uncle John, and some neighborhood kids on a chartered deep sea fishing boat. We'd jaunt out onto the Atlantic Ocean, often ten to twenty miles from shore.

On hot summer days, and with no break from the sun, we'd be out on that rocking boat for twelve hours. Add the thick smell of freshly caught tuna and blue fish, and you have the perfect recipe for a world-class case of seasickness. There wasn't enough ginger ale or Coca-Cola on the boat to settle my stomach.

My equilibrium has always been out of whack, and even sitting in the back seat of a moving car makes me nauseous. But not going on these fishing trips simply wasn't an option. My dad was more than intimidating on this. He loved these trips on account of his gregariousness and love of the sea. So I would spend most of the time belowdecks, throwing up and wishing I could be anywhere else but on that pitching boat.

My brothers used to laugh at me. The last thing you want as the youngest son is to attract ridicule from a boatload of males and be forever stamped as a milquetoast who couldn't tough it out. I was and am a thick-skinned fellow and not one to back away from a challenge. But this was about all my stomach and sense of balance could take. Still, I somehow made it through those miserable trips.

But that wasn't the worst of it. After our lurching and rolling at sea, my dad would always demand that we scale, gut, and clean our haul as the sun set in the late afternoon on the docks. We'd all be sitting in a slop of fish blood, guts, and scales. The stench was unbearable, and the memory of it still riles my nose hairs to this day. Whatever lunch I had been able to choke down during our trip was now a part of the waters around the docks.

It is from this history that I determined never to eat fish or set foot on a fishing boat again. The slightest waft of fish roils my stomach up into my throat in a knot of nausea. Hence, I don't eat in sushi restaurants, and no fresh fish ever finds its way into my refrigerator, much less my house. I even avoid walking by the fresh fish case at grocery stores, even upscale ones like Whole Foods.

And in my forty years in the restaurant business, I've never, ever composed a menu that included salmon—the most egregious fish stench culprit in the world! I have learned to tolerate most white fish—only saltwater white fish—and most shellfish. But never will you witness me putting raw fish into my mouth. Ever!

* * *

Fast-forward to January 2019. I'm in Tokyo for a meeting with executives from Toridoll Holdings Corporation (THC), a billion-dollar-plus global restaurant investment powerhouse that operates Marugame Udon Japanese Noodles & Tempura. THC is a publicly traded company that had over one thousand, five hundred restaurants worldwide at that time.

I was part of a team to discuss a proposal to partner with the company to expand the Marugame Udon brand across the United States. The group included Takaya Awata, founder, president, and CEO of THC. It was an immersion trip into Marugame Udon's history and culture, as well as an introduction to the THC brain trust to gain their support for this new partnership.

After a long, direct flight from Dallas to Tokyo—fifteen tedious hours in a jumbo jet draped in total darkness—we were whisked from the airport to our hotel. We were told we had exactly one hour to freshen up and ready ourselves for a formal dinner with Mr. Awata and his executive team. We were to meet at a very upscale and pricy restaurant: a favorite of Mr. Awata's.

Odds were high that this meeting would not go well. For one thing, in a stark departure from traditional formal business wear, I don't wear socks. Ever. They're restricting, and as my waistline has expanded over the years, I find them terribly inconvenient to put on. To this, add that we had little time to prepare for what promised to be a long night of omakase service (serve us what you'd like us to eat). I wasn't mentally prepared for omakase surprises.

I typically do not travel in groups. Over my years of consulting, I've been free to dictate how my days and nights pan out. Attending a twenty-person omakase dinner in a traditional Japanese restaurant where the menu was set, the table was horigotatsu—a very low table with a foot cavity—was a recipe for occupational catastrophe. After all, I'm six foot four and 275 pounds. So even getting down to a seating position was a critical test of composure.

As I struggled to sit, I noticed that I had been strategically seated directly across from Mr. Awata. Traditionally in Japan, the two highest ranking people sit at the middle of the table across from each other. So not only was I now completely uncomfortable, but I also realized I would have no margin for error with respect to my performance at this important meal.

I'm usually at no risk of alienating people with rude, inappropriate behavior when interacting personally or professionally. But to successfully sit through a methodically staged three-hour meal cum business meeting is nothing if not a grueling test of my patience. And if that wasn't enough, I was facing a fourteen-course meal featuring a myriad of sushi dishes. I was in grave danger of reliving the heaving glory of my fishing adventures off Long Beach Island in front of some of the most powerful players in the international restaurant business.

Mr. Awata spoke no English, and my Japanese is nonexistent, which only added to my predicament. And, of course, the first item brought to our table was fresh fish served sashimi-style, which, to my pedestrian fish mind, means a bigger slab of fish scent. And once that essence whacked me, I was immediately shuttled to those New Jersey docks carpeted in fish entrails. My stomach vapor locked.

Mr. Awata flashed his addictive smile and seemed anxious to know what this huge American thought of his favorite cuisine. He was well-briefed on my work in the restaurant business, and the fact that I was positioned directly across from him was a testament to his respect for me. I nervously explained to his interpreter that my long flight had left me with a sensitive stomach and that I would have to pass on the first few dishes until I recovered my sea legs.

It seemed like a natural enough excuse, and I believe I had successfully masked the true reason that I was passing on the food offerings. As I waited for what seemed like a transoceanic flight span of time for my words to be translated, I sensed he was perfectly okay with my sentiments. He held up his sake cup, indicating that at least I could drink.

"Kanpai!" he said loudly, lighting up his face with a grin.

Kanpai is Japanese for "cheers."

The great thing about sake is that it is often served in opaque wooden or ceramic cups, and no one can see how much you are drinking. "Kanpai" was my refuge, and the sake went down smoothly and often.

As the fresh fish courses dwindled in frequency, we were served cooked food that included tempura and a few grilled items. I ingested these with great relief. Maneuvering out of my down-low sitting position to use the restroom is a bit of choreography I will leave to your imagination. Let's just say this evening will be etched in my mind forever, and my hope is that it will never again be repeated.

* * *

I was a key player in a team of seven that included design guru Dana Foley; Brian Fauver, a partner in the restaurant investment firm Hargett Hunter; management whiz Pete Botonis from my Pei Wei days, and a few social media and marketing guys. We embarked on a one-week tour of the central and southern parts of Japan, with much time spent in the Kagawa prefecture, a Japanese administrative district on the island of Shikoku from whence udon was born.

Udon is a deliciously thick noodle—a Japanese comfort food made from wheat flour. It is prepared and served in a variety of ways. The most common is in a mild broth called kakejiru made from dashi, a stock from simmered kelp and fermented fish, soy sauce, and mirin, a rice wine. It is usually topped with scallions, prawn tempura, tofu, sliced fish cake, and spice.

We ate udon morning, noon, and night for days, occasionally interspersing meals of other cuisines before heading north from Marugame across the Kagawa region on a four-hour drive to Kobe. Eight of us were loaded in a basic passenger van with luggage, all stuffed in like ground sausage in a casing. I was awarded the front passenger seat. Our driver? A young chef from the Toridoll company who spoke no English and who answered every inquiry with a huge toothsome grin and sparkly, enthusiastic eyes.

His name was Aki, and I was informed by our interpreter that he lived in Tokyo and had never driven a van through the countryside. He was the only person available to drive us, and that was that. His inexperience was immediately evident as we barreled down the highway. Panic crept in with every chaotic lane change at seventy-five miles per hour and some herky-jerky swerving.

At one point, sensing that I was the only one paying attention to our plight, I turned to my traveling companions in the back.

"Stop fucking around back there," I barked. "This is some serious shit we're in."

After a moment of silence, the group howled with laughter. I was quickly given a huge bear hug from behind by the gregarious ringleader Pete Botonis.

"Don't worry, big fella. I got you," he said as the rest of the team guffawed again.

Aki joined in. I stared at him without a hint of a smile on my face.

"Please slow down and stop changing lanes so much!" I hissed.

Our interpreter translated. The cheap seats again hooted with laughter.

After two hours of white-knuckled riding, we pulled into a rest stop. Everyone piled out of the van, but I went directly to our interpreter.

"There's about $2 million in salaries in that van and at least ten family members back home in the United States hoping to see us all again," I yelled. "Tell Aki to slow down, stop swerving so much, and get us to Kobe in one fucking piece. Please!"

We got back in the van; Aki smiled broadly at me.

"I understand Mark-san," he nodded.

Our interpreter must have taught Aki a few words. The remainder of the ride was uneventful. My blood pressure normalized, and soon, the beauty of Kobe's evening skyline came into view. We were all ready for more food. And sake.

We spent one night in Kobe and dined at Mr. Awata's original restaurant just outside the city limits. It wasn't a Marugame Udon, and it featured a much broader menu, including skewers prepared on a robata grill, a Japanese charcoal apparatus used for cooking small sticks of proteins. Plumes of fragrant smoke billowed from those small grills, and the allure of charred meats filled our nostrils.

After days of nothing but noodles and dashi broth, I was ready for some substance. Over the next few hours, we devoured skewers of perfectly grilled steak, all parts of the chicken, and wonderfully blistered vegetables. My interest in robata began to percolate at that restaurant, and eventually, I would incorporate the grill into our new vision of Marugame Udon for the United States.

During this trip, my focus was on absorbing as much as I could about Japanese culture. After all, consumers in Japan were markedly different from those in Dallas, Denver, or Chicago. Our goal was to merge these culinary customs with a formula that would work in locations throughout the US—a tricky assignment.

How can we appeal to a broad-enough spectrum of consumers to grow in diverse demographic areas without losing our core? A concept like Marugame Udon doesn't have a broad field of competitors because ethnic cuisine by nature has limited market potential. There are seventy-two McDonald's locations in Los Angeles, for instance. But that same market might support only six or seven Marugame Udon locations—even though the prices might be similar.

So our goal was to determine just how far we could stretch the menu to widen the appeal without jeopardizing the concept and the expectations of our Japanese partners. This pressure is hard to explain because we are talking about investment dollars, the reputation of a well-established publicly traded company, and the hopes of investors. Much of this pressure fell directly upon my shoulders.

But I felt 100 percent confident we could achieve our goals. Once we were back on US soil, we could materialize our vision and turn Marugame Udon into a nationally recognized brand. The first few months of our work went incredibly well. We began on the drawing board, turning our vision into a new prototype while taking the reins of the existing units in California and Hawaii, putting them on track for improvement.

The entire plan was presented to THC, and the deal was sealed. Marugame Udon USA LLC was formed, and we began to take the reins. I was appointed president and CEO of this new venture. When interviewed for this position, I described my role with a metaphor. I was playing the role of an experienced tailor. THC had come to us with a beautiful suit that needed some alterations.

It was gorgeous material, but it would not fit the new owners. The look and feel needed to be modernized without changing the overall style. Subtle alterations made it more contemporary and appropriate for the occasions for which it would be worn. That was what I was doing: altering Marugame Udon to better suit the broader US market. We all agreed this metaphor made perfect sense.

* * *

For those unfamiliar with private equity and how it applies to growing restaurant companies, here is a simple summary. Private equity firms raise funds, or PE funds, to invest in companies they believe have great potential for growth, regardless of the industry. There were about eight thousand private equity firms in the United States as of 2019, and each one might specialize in different industries.

The largest invest billions of dollars in potentially profitable companies that might just need a financial boost to attain new levels of growth. While these companies generally invest in promising brands, they also back failing companies that haven't reached their potential or who possess strong brands in decline.

In the restaurant industry, the top PE firms typically go after successful fledgling brands that have reached just a small percentage of their growth potential. Some of the PE firms include Karp Reilly with The Habit Burger Grill and Sprinkles Bakery & Cupcakes; L Catterton Partners with Velvet Taco, P.F. Chang's, and First Watch Daytime Cafe; and Roark Capital Group with recognized brands like Auntie Anne's, Jimmy Johns, and Moe's Southwestern Grill.

In 2018, I was contacted by Jeff Brock, a principle of Hargett Hunter based in Raleigh, North Carolina. His request? To help with work on the fledgling Hawaii and California restaurant chain Marugame Udon owned by THC. Though the company had introduced Marugame Udon in Waikiki and had launched subsequent locations in California, THC had not made another investment in the US for several years.

But by 2016, they figured it was time to expand Marugame Udon into the US mainland, for which they set up an Irvine, California-based team. With more than one thousand units worldwide, Marugame focused on freshly made udon noodle bowls and tempura in a cafeteria-style serving mode. But after launching eight units over two years on mainland US (in addition to the one in Honolulu), bosses in Tokyo determined the US team was underperforming expectations.

While the Hawaii location is enormously successful, with a constant flow of tourists, many of which are Japanese, the mainland US consumer was a tougher nut to crack. Of the eight stores operating in California, half were wildly profitable, while the other half were sluggish. THC decided it needed outside help. They solicited Hargett Hunter. Jeff was someone I had met years earlier during my days with Pei Wei Asian Diner. We'd struck up a quick relationship because we shared a love of basketball, restaurant concept development, and great food.

"Mark, I finally have a project where your experience with Pei Wei can help us," he said. "It's an awesome opportunity for us to get a foothold with an Asian national brand in the US. You're the perfect guy to lead the charge."

As I was still doing my consulting work and had the time, I was thrilled to be a part of this assessment team. Jeff also secured the assistance of associates in his own firm, and an additional team of six of us set out to develop a detailed report on Marugame Udon and its potential as a national brand in the US. It was a consultant's dream job.

The challenge was determining what to do with a brand that had such a puzzlingly schizophrenic impact on the fast-casual Japanese dining market. In California, it was primarily located in areas densely populated with Asians. But about half of their units were in areas seemingly without any strategic rhyme or reason. Some were free-standing. Others were wedged in malls near food courts. The remainder were in cities without a street presence or available parking.

But the other half were home runs. They had long lines, eye-popping sales, and frenetic followings. Because of my work in Waikiki, I was familiar with Marugame because it was thirty yards across the street from Michael Mina's food hall. I'd eaten there a few times, well before my phone call from Jeff. So I had a head start on my part of the assessment. My part of the project targeted the food quality, dining experience, and store-level organization and operations.

Under the Hargett Hunter banner, we would produce a one hundred-page report full of charts, graphs, recommendations, and a dizzying array of financial statements and competitive analyses. PE firms like Hargett are information junkies. My own part of the report came to 20 percent of the meat, but the team was anticipating my qualitative assessment with a great deal of interest.

I provided a mostly glowing report on the qualitative aspects of Marugame Udon, giving the concept high marks for quality and value, medium grades for convenience and dining experience, and a low score on team hospitality. I recommended Americanizing some menu offerings by adding a teriyaki platform, elevating the ambiance by improving lighting and adding more comfortable seating, and improving management interactions with guests. We did not believe that a menu consisting primarily of udon (a noodle relatively unknown to Americans) and tempura would cut it in cities outside the coasts.

The Toridoll team in California—Toridoll Dining California (TDCA)—was directly in the crosshairs of our report. We quickly saw why the parent company in Tokyo was disenchanted with the team. There was far too much management turnover, far too little information sharing, and little depth at key levels. It was clear that TDCA didn't trust us or want us around. But as uncomfortable as that was, we strived to be candid, to understand operations, and to rigorously analyze the information shared with us.

The Toridoll team in Japan made it clear that they considered the US market fertile ground for expansion. And our report did nothing to diminish their enthusiasm. The next question was, "Who was going to grow Marugame Udon in the US?" At a dinner in San Francisco at Michael Mina's Japanese-style pub, Pabu, I sat with Jeff and two of his associates, Brian Fauver and Trey Stilley. Our discussion turned to the future of Marugame Udon.

"Why not us, Jeff?" I asked. "Why don't we figure out how to partner with Toridoll and become their operations arm in the US? I can assemble a team to replace the TDCA, we can train and upgrade the current management teams, and we can put a new prototype in play to show our vision."

It was clear to everyone at the table that I was itching to grow another concept. My work with expanding fledgling concepts was solid, and Jeff and his team had access to the capital to do it. A plan was hatched over cocktails, skewers, and Mina's Japanese cuisine. THC had designs on being among the top ten restaurant companies in the world by 2025. It was teed up perfectly.

***

It took a long and painful six months to negotiate the deal, but eventually, it was solidified. We formed Marugame Udon USA, LLC to work as a minority partner. THC had the rights to buy into existing Marugame Udon restaurants in the US, while we secured the exclusive rights to expand the concept in all fifty states.

We developed a value creation plan, spelling out a five-year strategy for our new partnership. Once the ink was dry on our agreement, we had a 20 percent stake in Marugame Udon, with THC retaining the right to purchase back our share at the end of the five-year term. Between franchising and company-owned stores, we agreed to open ninety restaurants by 2024.

To help put the plan in place, I reached out to fellow "restaurant whisperer" Pete Botonis.

"Pete, do you have time to listen to an idea that I have?" I asked. "It could be an opportunity for us to catch lightning in a bottle one more time with Asian food."

Pete was key to our Pei Wei expansion years before and the most solid, respected, and capable operations guy I have ever known. He had just disconnected from an entrepreneurial venture in Chicago and was available. By offering a nice salary, an ownership stake with a good upside in five years, and a great concept to build, I was able to convince him to come on board.

For the first six months, our efforts in this project went incredibly well. We structured our vision for changes into a new prototype while taking the reins of the Marugame units in California and Hawaii to build those restaurant teams and create more productive cultures.

Pete was immediately embraced by the management teams, and our group's hands-on approach to operations was welcomed by restaurant staffs. We were approachable, engaged, and not afraid to work long, hard hours—something the teams weren't used to with the previous leadership. I game-planned menu changes and more impactful restaurant design. I had a strong base of managers on board committed to executing our concept revisions. And then, in March 2020, disaster struck: the COVID-19 pandemic.

In the first quarter of 2020—a period we now call pre-COVID—Marugame Udon's performance was stellar. We exceeded budgeted revenues and projected profits and were riding high with a new sense of confidence. The prototype was coming together, we were developing a rhythm as an executive team, and we were earning the confidence of our Japanese partners.

The first few calls concerning this new health crisis caused alarm but not panic. Available information was sketchy, news reports were inconclusive, and daily information was confusing. Phone lines started buzzing about what impact COVID-19 would have on our business. At first, our days were occupied with dozens of phone calls discussing what kind of disruption would impact our business. Would we have to shut down? Would we have time to re-forecast our budget? Would we be able to travel to our restaurants?

There were no solid answers. More troubling was that the Marugame service style was cafeteria-line driven. Guests selected their own dining options, with udon bowls built to order. We were not set up for delivery or to-go service, and our point-of-sale system was not integrated with delivery apps. Some of our locations were "mall-locked" without exterior signage. They were positioned to take advantage of shoppers and strong flows of walk-by traffic.

But as malls closed, our Marugame mall locations shut down too. In March 2020, at the start of the pandemic, our first Texas prototype was on the drawing board. It was designed to dramatically increase our dining takeaway capacity. But with the suddenness of the pandemic, our existing stores were badly exposed.

We realized we were at high risk of being shut down. At first, it seemed like we'd be closed for maybe a month or two while a solution was found. But as anyone reading these words knows, the restaurant business—especially those concepts not designed for takeaway business—was devastated by closures and restrictions.

The impact of local regulations hindered us from moving forward. No two local governments assessed or reacted to the situation in the same way. In California—the state driving 80 percent of our sales—decisions were left to county governments, and solutions varied. Each county with Marugame locations—Orange, Los Angeles, and Almeda—had their own restrictive regulations in place. There was no consistency in local government messaging, and we struggled to make sense of what we could and couldn't do. They mostly determined that no one would be allowed to eat in restaurants for the foreseeable future.

Yet as important as the restaurant business was to our economy, local governments permitted a varying mix of delivery, patio dining, and takeaway service. Our entire business was thrown into turmoil in a sea of uncertainty. No one in our business was prepared for this level of disruption, and panic set in. Having no revenue while carrying labor and rent costs will do that.

When it was apparent our business needed drastic adjustments, I planned a "COVID-19 Summit Meeting" in Los Angeles at a shutdown Marugame in the Westside district of Sawtelle. My flight into LA had maybe ten people on board when it arrived at an apocalyptically quiet LAX in late April 2020. I took videos of the terminal and sent them to our team, illustrating how frozen in place Los Angeles had become.

At the summit, we outlined a plan to shift our focus from dine-in to take-out and delivery. Our strategy consisted of scrapping our entire POS system, ordering new phone lines, and implementing better food packaging and measures to operate with no guests within our doors. What had been about 5 percent of our business needed to become 100 percent if we were to survive.

When we first took the reins pre-COVID-19, and I was relatively unknown amongst the teams, I walked into a Marugame location in the Beverly Center mall in central Los Angeles to place a big order for my office for the following day. It was lunchtime, and there was a throng of people with a line out the door and a bustle of activity. Takeaway was the key in my view, and I wanted to know more about how we handled it.

"We don't do to-go orders," I was told. "You'll just have to come in, get in line, and place your order that way."

Achilles heel times ten. Turning away business like this is borderline arrogant, but Marugame had gotten away with it for years. There was no portal to place orders on our website, no catering manager, and no to-go order forms to fill out. This weakness became a huge fucking flaw under the glare of COVID-19. Sales plummeted, and we worked frantically to shift our core business. But it was like bailing out a sinking ship with an ice bucket.

We did manage to open two new stores with highly efficient takeaway operations in Dallas during the pandemic—Texas's government was far more business-friendly than that in California. But after strong openings, sales slipped. People hadn't returned to offices in any significant numbers, restaurants were plagued by labor shortages, and the public was not dining out in the numbers they were pre-pandemic.

If there's not enough revenue and no relief in sight in the short term, heads roll. I received a call on a Friday afternoon in Dallas and was asked to step down as president of Marugame Udon USA after eighteen months at the helm. Brian Fauver, my direct contact at

Hargett Hunter, and I had our differences over the time we worked together, but I wasn't bitter over the end of my short reign.

The only other time in my career that I'd been fired was at Canyon Café, and I was reinstated six hours after termination. That wouldn't be the case with Marugame. The end came down to the fact that we couldn't shift from a dine-in-dependent business swiftly enough. The reality is that I'm not very good at running someone else's company. My frustration with administrative requirements and meetings was my downfall once again.

While I love in-the-trenches teamwork, I've always known I don't suffer people in positions who either think they know better or who don't understand the rhythms of our business. You can't sit in front of a computer and make judgments based solely on figures, spreadsheets, and projections. But Marugame was hemorrhaging cash, and the partnership was unified in its decision to remove me.

Marugame—more so than most restaurants during the pandemic—suffered from inherent weaknesses. Time will tell how new leadership adjusts. I have a shooter's mentality: I don't punish myself for the shots I miss. Trust your aim, experience, and mental approach, and get ready for the next shot. That shot was Bizzy, a burger concept I was finally ready to zero in on after years of shoving it aside.

16

Keeping Bizzy

For the first fifteen years of my life in New Jersey, our Sunday routine was to go to church and make a trip in our station wagon full of family to either Wildt's delicatessen or the closest White Castle burger joint in Verona. I remember back in those days of limited fast-food choices, White Castle was something we all craved. We'd buy them by the sack. Steam laced with traces of griddled onion, burger meat, and fresh rolls would fill our car with narcotic aromas.

It was pure bedlam once we got home with an over-stimulated family of kids clamoring for our rightful share of burgers. There was never a morsel left when we'd all finished devouring the contents of that sack. It was a feeding frenzy in our smallish kitchen that I am sure earned my parents a midday cocktail or two. Yet as quickly as the tsunami of eating passed, the smell lingered in the air. White Castle was my first real food obsession.

Though my move to Texas eventually took me out of range of White Castle expansion plans, I was able to travel often enough to the east and Midwest to fulfill my ongoing obsession with those tiny square slider cheeseburgers. As I write this today, you can find frozen White Castles in most supermarkets, and their tenacles have taken them to once-inconceivable locations like Scottsdale, Arizona, and Las Vegas.

Stories of these openings are legendary. Lines snaked around blocks and buildings, and supplies quickly vaporized. Displaced easterners and Midwesterners stormed the new locations, and new territories thrived. White Castle was the East Coast and Midwest cult equivalent of In-N-Out Burger on the West Coast.

The difference? White Castle specialized in mini burgers, and In-N-Out Burger was the more traditional hamburger with conventional toppings. No White Castle slider has ever been visited upon by cold lettuce or sliced tomato. White Castles were limited to chopped pan-griddled white onion and tiny warm dill pickle disks. In my mind, White Castle had it right. Why put something cold on a hot burger? Why confuse your palate with competing temperatures and textures?

Legend has it that the term "slider" was coined by an inebriated devotee who felt that consuming a White Castle hamburger or cheeseburger was easy because the burgers were small and greasy enough to slide down your throat without chewing. Makes perfect sense. Late night trips to White Castle—many open twenty-four hours to satisfy the late-night cravings of the smashed and plastered—were simply part of White Castle mania.

The cult feature film *Harold & Kumar Go to White Castle*, a 2004 buddy comedy, chronicles a couple of stoners watching a White Castle commercial before setting out to find one in their zonked state. Many episodes of the CBS sitcom *King of Queens*, which ran for nine seasons starting in 1998, point out that White Castle is lead character Doug Heffernan's favorite food. The White Castle logo is easily and quickly recognized.

And while White Castle's growth over the years has paled in comparison to industry giants like McDonald's, Burger King, and

Wendy's, White Castle, at 349 units (compared to McDonald's 40,031 locations, making it less than 1 percent of McD's size) remains privately owned. It also can proudly claim *Time* magazine's 2014 title as "the most influential burger of all time." Its impact on the American burger scene is undeniable.

In 2014, I decided to focus my creative energy on this unwavering childhood obsession. And while burgerdom is a crowded space dominated by hungry Goliaths, I felt that, in Texas, there was a void in the smaller, "better burger" market. I'd enjoyed resounding success with fast-casual restaurants. So this was right up my alley.

After some noodling around, travel, and some quiet time, I ended up with Bizzy Burger. Don't ask me how I came upon the name Bizzy. (I do have a niece named Elizabeth, shortened to Liz, and even Biz—but never Bizzy.) The name had clever applications ("Let's Get Bizzy" was a t-shirt and rallying cry), and my goal was to create the adult version of White Castle for the not necessarily inebriated. My aspirations are elevated, I know.

Sometimes it's just that easy, nothing more than an intuitive moment. A thought bubble. A name comes out. I also began to think in terms of becoming "the electric car of the burger industry," and that type of thinking drove me down a path that I didn't see many others driving on. I wanted to be a disrupter in a sea of similarity. Bizzy is a smaller burger with an infinitely crave-able platform of offerings.

To be a disrupter, you must be committed to disruption as an action, not just a slogan. The fast-food industry had become too much smoke and mirrors, reshuffling a deck with the same old fifty-two cards—not innovating, just aggravating with tedious sameness, new names, and "meal deals."

If Bizzy succeeds as a franchisable model, franchisees must commit to an exhibition greenhouse to grow herbs for our hot proprietary "Blendz," like flame-seared hot pepper concentrate, blistered tomatoes, herbs, and roasted wild mushroom puree. We offer guests these concentrated culinary concoctions to top their sizzling burgers. Franchisees must also commit to sourcing locally produced Akaushi beef and King's Hawaiian Original Hawaiian Sweet Rolls.

We'll also require them to hire a local artist to paint a one-of-a-kind mural emphasizing local culture inside or outside the premises with phrases that poke fun at other burger joints. These include "Lettuce is a lie!" "Tomatoes are imposters!" "Avocad-be-kidding me!" "Pickles on placate!" "Raw onions = Raw Deal!"

There will be no cold toppings on Bizzy Burgers. You wouldn't do that to your steak, would you? We also feature a veggie sandwich with vegetables that you can identify and are void of the typical rice, bean, or quinoa extenders.

Since I traveled often to Scottsdale and occasionally to Las Vegas, I continue to analyze the attraction of White Castle. I was never convinced my own vision might be off base. Bizzy had a solid place in the crowded burger domain, and I set out to see who agreed with me.

I developed an effective presentation and held a few tastings. After raising funds for Bizzy back in 2014 and assembling a team to help bring it to fruition, the investment fizzled when some of the partners bailed out for various reasons. I abandoned the project and put the presentation into my proposed concepts file called, appropriately enough, "the vault."

Seven years later, after my stint at Marugame Udon ended, I took a serious look in the mirror.

"What makes you happiest, Mark?"

"Creating something new," was my answer.

I'd felt I performed well with Marugame Udon. But the truth is that what I had envisioned for Marugame wasn't the right fit for that partnership. My role was essentially akin to being a franchisee of a very large brand. And while I readily admit that my vision for Marugame differed from that of our Japanese partners, only time will tell how that will turn out. I was ready to get back in the game to what I do best. I was ready to create. Clean canvas, new paint, uncluttered mind.

* * *

One of my close colleagues since my days with Brinker was Chili's founder Larry Lavine. Larry and I both traveled the same unpaved road of conceptual development. He loved tinkering with concepts, bringing new ideas to the table, and investing time in the "creative test kitchen of life," or the mad scientist laboratory. In 2021, Larry was once again restless for something new. He'd had his hand in multiple concepts, but lightning rarely strikes twice in the same place.

He'd tried Marugame Udon—as he did most of the concepts I've presented in this book—and was a big fan. He asked me if I had anything else that he and I could work on together. We met on a Saturday morning at Eatzi's (three of my previous concepts—Eatzi's, Velvet Taco, and Pei Wei—were all crammed in two different corners a mile from my house). We chatted about new concepts and our opinions about them and eventually got around to discussing Bizzy Burger.

I shared that I probably wasn't going to get another executive position, and our discussions about Bizzy became front burner. Being coconspirators in something new was probably the best and most natural option for both of us. Our personalities didn't compete, our palates were similar, and our combined histories on the Dallas restaurant scene would be hard for anyone to ignore.

"You know, Larry, neither of us are getting any younger," I said, realizing he was seventy-two, and I was sixty-eight. "Da Vinci painted his masterpiece the *Mona Lisa* in his late fifties, so we're not that much older to create ours."

For the next hour or so, we had an animated conversation that passersby might have mistaken for two old farts arguing over an upcoming Cowboys-Redskins game. Instead, our raised, excited voices were focused on the rebirth of the next big thing in fast food.

"Let's do this, Larry," I said. And we shook hands.

Larry, as the founder of the first full-service, casual restaurant burger chain, and I, who was known in Dallas as the "father of Dallas fast-casual," were a formidable team. And both of us were available to spend the time to bring Bizzy to life. All we needed was a chef. John Franke of Velvet Taco fame was now a free agent and had started a wildly successful culinary consulting company.

He'd left Front Burner Restaurants after cashing out on the sale of Twin Peaks. He decided he'd had enough of the corporate chef life. He was building a family, had cash in the bank, and wanted a life more suited to being a dad. The timing was perfect, and I acted quickly to convince him to join the Bizzy project. In exchange for his help developing Bizzy's very limited but chef-driven menu, I offered him a small fee and a piece of the pie.

What Bizzy needed was another "Mark and John" like that of Velvet Taco. After some skepticism, John decided that he'd come along. Larry, John, and I met, talked, debated, laughed, and shared. Our combined experience was as impressive as any threesome in our local industry. We set out to raise $1 million and called on friends, investors in past projects, and friends of friends.

It took about three months, and we had fully subscribed our capital goal. Along the way, I added an ex-investment banker to the mix, and the team was now as complete as any I've ever assembled. Soon after all the legal work was put to bed and the documents signed, we started having tastings at my house. My kitchen is spacious but abysmally vented—more designed for eggs and bacon breakfasts than an array of three cast iron skillets smoking away with ten to twelve sizzling burgers at a time.

After each tasting, I'd smell charred burgers in my house for weeks. But seeing the raised eyebrows on our investors' faces and listening to their praises of John's work on the griddles made it all worthwhile. In between refreshing frozen bloody Mary's and Thai Coco Tea freezes, we'd gulp down Bizzy burgers dripping with juice and slathered with hot Blendz. I smiled and answered questions.

"How quickly can we find a place and open up?" was the most frequently asked.

It didn't take long to find a great site for the first Bizzy in North Dallas, just a mile from my home. As I write this, I have signed off on a killer design presentation, and our plans are being reviewed by the City of Dallas. An October 2022 opening is our goal. As we prepared for that opening, another opportunity came up.

Credit again to my good friend Derek Missimo. He and his partner had two restaurant operations in Phoenix Sky Harbor Airport in partnership with his pals at airport restaurant developer SSP America. One of those restaurants, Smashburger, had closed during the pandemic. After a series of conversations, Derek decided not to reopen that space. The City of Phoenix, the landlord of all airport properties, had issued Derek a timeline to either reopen the space or find a buyer who would. Derek called me.

"Mark, I have a proposition for you," he said. "How would you like to open a Bizzy in the Sky Harbor Airport space where we had Smashburger?"

I was familiar with the space with my regular travels in and out of Phoenix and had often suggested that Derek give me a shot at the space when they chose to reopen it.

"You're fucking kidding me, right?" I said. "Is that even a question?"

I was all in on the chance to prove Bizzy as a successful airport concept and to offer another opportunity for my investors. It took about a week to convince the Bizzy board of directors—a group I headed and lovingly called Board Bizzy—to approve the capital raise.

"My friend, you have a new tenant!" I said to Derek excitedly.

And before we had our construction plans back from the City of Dallas for our first Dallas location, the Phoenix market looked to be the new frontrunner to bring Bizzy to life!

An intoxicating blend of emotions. That's what a new concept is. Here we go again!

TO BE CONTINUED!

17

The Vault

You have now read about the restaurant concepts and projects I've been involved in over the course of my career. But not all my work or the clients I have collaborated with matriculated into the restaurant world—far from it. Much of this creative work was fully developed on paper, depicted in renderings and kitchen design, and had menus that were minor tweaks away from showtime.

I've collected these ideas and stashed them away in an area I have dubbed *The Vault*. Call them short stories if you want, but they have a purpose. Maybe, someday, they'll be released from The Vault, dusted off, and adopted by someone who sees the possibilities and is willing to risk bringing them to life. Here's a synopsis of just a few of those, starting with a project called Pinky Chan.

PINKY CHAN

Just writing the name makes me smile. Pinky Chan is an Asian restaurant built around a fictitious and mysterious Chinese chef who grew up in an orphanage and found her way to the restaurant business after a whirlwind journey through southeast Asia. The concept was first brought to my attention by fellow restaurateur Mico Rodriguez. Like me, Mico is a serial concept developer who is always thinking about the next creation.

His track record is legendary in Dallas. But his palate is anything but local. On the menu at Doce Mesas, his Mexican restaurant, he pays homage to his love for Asian cuisine with Chin Chin Chicken, his version of the famous Chin Chin chicken salad from the restaurant of the same name in Los Angeles.

And while he was busy with Mesero, his upscale Tex-Mex concept in and around Dallas, he had an itch for a new concept he asked me to scratch called Pinky Chan. It had a vague focus, but I was immediately drawn in by his vision and was committed to bringing it to life. But he didn't have the time to dedicate to Pinky Chan's development, though his fingerprints were all over it.

I called on my friend and graphics savant Amber Brown to assist. All we had to work with was Mico's story about his fictitious heroine on a journey that landed her in Paris as an acclaimed and renowned international chef. A good start. Amber skillfully interpreted the story and created a logo that was an effective visual expression. Mico enthusiastically endorsed the logo, and I set out to work on the menu. The simple Asian menu I composed was probably the most creative I have ever come up with.

I'd never created a menu that didn't follow the familiar format of appetizers, soups and salads, entrées, sides, and desserts. But this story and Mico's own sharing of his vision took me down a new path. I came up with an eclectic menu that categorized foods, not by meal course, but by how we suggested people eat that food. The categories were hands, fingers, chopsticks, and spoons.

Fingers contained foods like sticky wings and five-spice ribs, while Hands had wraps and some innovative sandwiches on scallion pancakes. Chopsticks had rice and noodle bowls, and Spoons offered Asian broths like Tom Yum and Pho. The menu was a big hit with our team.

We had no investors yet. In fact, it is rare that someone will commit hundreds of thousands of dollars or more to something they can't evaluate visually. So we set out to create a one-of-a-kind presentation package—a box. After opening the lid, investors were treated to a series of placards depicting our vision for this unique Asian restaurant. It was stunning. Mico immediately began ticking off the names of potential investors with whom we should share it.

Eventually, word of Pinky Chan spread to my partner and savior of Banh Shop, Derek Missimo. As luck would have it, Derek and his partner at Dallas-Fort Worth International Airport were bidding on a restaurant space in Terminal D right across from our original Banh Shop. I suggested Pinky Chan, and the wheels were put in motion. With so much of the development completed, all we had to do was craft a bid around the menu and the story.

The airport bidding process is notoriously long and complicated, with more holes in it than a colander. But if you want to do business at the airport, you must toe the line. Derek and his team had the patience for the process because they knew the reward—captive audiences, light competition, and all-year-long business.

Our bid had to include several tastings for DFW airport cognoscenti, and for that, I enlisted two of my favorite chefs: Uno Immanivong and Teiichi Sakurai. Uno was a pivotal character from my Trinity Groves days. Teiichi is the owner of the most-acclaimed Japanese restaurant in Dallas: Tei An. Classy and unassailable, Teiichi only came on board because of his connection to Mico. His commitment allowed us to use his restaurant for the tastings we'd have to conduct. I didn't have a restaurant at the time, and Uno was slowly phasing out Chino at Trinity Groves.

We held three tastings for the airport concessions team and their invitees in the Tei An private dining room. Each meal was an orchestrated symphony, impeccably served. I doubt the airport team had ever experienced anything close to this culinary opus. We fawned over them to the point of excess at times. Derek and his partner, along with Teiichi, invested tens of thousands of dollars in the bidding process. But that's what it takes.

In the world of airport food service bidding, the process is rarely about which team has the best proposal, food, or pedigree. Behind-the-scenes machinations count heavily, and I could tell that Derek felt we were a long shot. Forget that our food was flawless for three straight tastings. Discount the fact that the assembled players were a veritable dream team. Overlook our deep experience in airport concessions.

Pinky Chan didn't make the cut. The air escaped from our balloon overnight. The winner in this bidding war was a local group of well-known and accomplished restaurateurs who operated Shinsei, an upscale Asian concept. And though they won the bid, it was never embraced by the international travelers passing through Terminal D. I shook my head every time I traveled out of that terminal and saw mostly empty seats in the restaurant.

Pinky Chan would have been a better fit, but that hardly matters in airport bidding wars. Shinsei closed after just a couple of years in operation and was replaced with a Tex-Mex concept. I bring out the Pinky menu on occasion and share it with friends and colleagues, using it as a benchmark for what I believe is what innovation is all about. Reactions are always very positive, but Pinky is always returned to The Vault. She has aged well, and her light may yet shine through.

MARUGAME SCHOOL OF UDON

Not everything in my vault is a concept ready to be hatched. One of the most innovative ideas I came up with was a brand extension for

Marugame Udon that I called School of Udon. Udon was uber-popular in California and Hawaii but was being met with ambivalence in Dallas after we launched two locations. We were struggling with where to expand the concept, and we needed to present revenue-generating opportunities to our board.

I had to think out of the box on strategies to build revenue and, at the same time, elevate our brand and help it be more accessible in markets that weren't as familiar with udon. I attacked the challenge using a teaching platform. It was an idea inspired by my work in Honolulu at International Marketplace with the MINA Group and Billy Taubman.

Our Marugame in Waikiki was hands-down the brightest star in our constellation: It attracted one thousand, five hundred people a day and was located right across the street from the International Marketplace. On the top level of the mall were a few restaurant spaces that had never been developed but were perfect for a bold idea like my School of Udon. As CEO of the company, I met with the mall management to begin assessing the viability of developing a space for udon-making classes. I believed it would attract tourists for family activities and expand our brand image.

In addition, I saw this as an opportunity to create a universal training school for the growing franchise base of Toridoll Holdings, Marugame's parent, and to have a central facility for yearly outings, presentations, and board meetings—all in the paradise that is Waikiki. Working with Alonzo Cudd, my chief marketing officer, we composed an exhaustive twelve-page PowerPoint deck to present to the board and Toridoll Japan. I enlisted my design colleague Dana Foley, who had helped reinvent the Marugame menu, to create artist renderings of the School of Udon.

Her work included gorgeous views of a space with a merchandise counter for Marugame swag and a fully equipped double-sided exhibition kitchen for a class of up to two hundred people at a time.

The economics were unassailable too. The School of Udon would create revenue and launch our brand onto another attention-grabbing platform. The presentation was performed over a Zoom call. We were psyched!

But it went over with a lead dirigible thud. Our enthusiasm for the stunning visuals and its positive financial impact failed to make a ripple in the Pacific Ocean that divided us. In retrospect, this experience was the final burst of propellent driving me out of Marugame Udon USA LLC. The box I was forced to think in became suffocating. My vault grew larger by one idea.

BLEUBIRD. FRENCH 101.

Christophe Poirier and I sat on the sidewalk in the shadow of the Arc de Triomphe in Paris. We'd arrived at Charles de Gaulle Airport hours earlier after a ten-hour nonstop flight from Dallas. We'd taken a few minutes to check into our hotel, just steps off the famed Champs-Élysées.

Our mission? To visit the best Parisian sidewalk cafés, dine on croissants and endless servings of French onion soup gratiné, and game-plan a concept that could compete with the growing number of sandwich shops, all-day bakeries, and cafés in the US.

Wherever you are, you have at least one of these in your neighborhood—Panera, Corner Bakery, La Madeleine, Mendocino Farms, Au Bon Pain, and a range of others. Aimed squarely at the breakfast and lunch market, these concepts are designed for on-the-go dining and casual breakfast and lunch meetings. It was fertile ground for Yum! Emerging Brands to explore and, with Banh Shop and Super Chix firmly in development, its next promising target market.

Yum! Brands had a self-diagnosed blind spot in this segment. The newly formed Emerging Brands division, headed by Christophe and Scott Bergren, had this type of concept on their development short list to present to the Yum! executive team. This was a passion of Christophe's, and his enthusiasm for sharing his beloved Paris was evident from the start.

The list of places we needed to experience in Paris was long. I didn't complain. I'd always included the classic onion soup gratiné on my last supper menu, and I had enough energy to try as much of this food as we could consume during our four-day Parisian stint.

Our first visit was to one of the oldest bakery chains in France and one that had grown methodically and steadily since its founding in the early 1900s. Simply called PAUL, it offered a wide variety of sweet and savory pastries, sandwiches, macarons, and sumptuous Parisian coffees. In the heart of the Champs-Élysées, we noshed on croissant sandwiches as we relaxed in the slightly brisk fall afternoon.

If you've ever wondered what all the joyous hubbub is concerning Paris, spend an afternoon relaxing at a sidewalk café, taking in French cuisine and history in the famed Eighth Arrondissement, the district on the right bank of the river Seine and centered on the Champs-Élysées. The answer will arrive quickly and resoundingly.

On our list was a mix of chain and mom-and-pop bakeries. Destinations included the acclaimed Eric Kayser bakery (which was readying for US expansion, a move that would ultimately fail), Les Deux Magots (which translates to "the two Chinese figurines"), Café De Flore, and a diversionary dinner at one of Christophe's favorites, L'Atelier de Joël Robuchon, in the shadow of the Arc.

Our goal was not to replicate a Parisian street café on US soil, but to bring this classic fare and a touch of the environs to America in a way that no one else was doing—even La Madeleine Bakery and Café based in Dallas. We felt we could create authentic French in an affordable, accessible setting that elevated café dining in the US.

We departed the City of Lights with full bellies, minds teeming with ideas, and overflowing bags of Parisian baked goods and even perfumes (I had plans…). The flight home gave me a head start on the project. I never sleep on planes, and over the course of what amounted to a full day's work, I strategized our next steps.

These included connecting with Plan B Group to design a prototype, and Chef Tom Fleming, who owned Crossroads Diner, to help develop the menu. Crossroads Diner served just breakfast and lunch, and Tom had adopted this simpler culinary lifestyle so that he could spend more time with his family. He was trained in classical French cuisine and had spent a fair amount of time in France honing his skills with some well-known classical French chefs, such as Paul Bocuse.

With Tom doing the food and Plan B doing the design and graphics, we had a team more than capable of executing the concept and my vision for the menu. After a few months, we conducted a series of tastings at Crossroads Diner with American takes on an onion soup sandwich, niçoise salad, croque monsieur, steak frites, and canelés, a French pastry. They were presented in a format that was familiar, cost-effective, and irresistible.

Plan B, under the direction of Royce Ring and Alex Urrunaga (from my Velvet Taco experience), created a dynamite design rendering with the name Bleubird. It was easy to pronounce, radiated with feminine appeal, and was far more attractive than anything on the market. Roger Bergren and Christophe were enamored with Bleubird and the caliber of the brand package. All we needed was a nod from the Yum! executive team.

But it wasn't to be.

I can't tell you exactly why. I was only a consultant. But I can guess. Yum! has executable brands that are models of efficiency and easily franchised modes that don't translate well with sophisticated venues and diners. French food was considered fringe in what I am sure they considered a niche market.

And while everyone on our team felt we hit the bull's-eye in a growing market, Bleubird just didn't resonate with Yum! decision-makers. In my opinion, Bleubird was the lone Emerging Brands concept with the greatest potential impact—both nationally and internationally. French 101.

More than half of the world's population is female. And while I have no quarrel with Yum!'s core brands and the success they generate, there isn't a single concept in that stable tailored to appeal to women. Take it for what it is. Or leave it in The Vault.

Epilogue

It was a brutally cold, snowy winter day in Ithaca, New York, on February 12, 1973. I was living in the West Tower dormitory on the Ithaca College campus. Most of the classes that day had been suspended, not so much because students couldn't walk to class, but because inclement weather prevented campus workers and professors from reaching the school. Ithaca College sat on the south hill of Ithaca, and the long trek up that hill was notoriously dangerous in weather like this—which seemed to be most of the time between November and April.

Most of the thirty guys who resided on the sixth floor were lounging in either the dorm lobby or their rooms. Room doors were open, and we all moved freely around the floor—a close-knit group. My roommate Steve Perez and I had a corner suite on the opposite side of the hallway, and that hall space had the lone pay phone on the floor. We never heard it ring, but it was our only connection to the rest of the world—long before smartphones.

There was a knock on our door frame.

"Breeze, you have a call on the payphone," said my friend Jake Kennedy.

Jake was one of my best friends, and we played basketball together on the Ithaca College team. Those guys nicknamed me Breeze—short for Breezinski. I got up and made my way through the hallways to the pay phone. The receiver was dangling from the cord. There was no chair to sit on. I reached down and picked it up.

"Hello?" I spoke.

"Mark, it's your father," said the voice on the other end.

This was odd. My dad had never called me over the three years I'd been away at school, though my mother would check on me occasionally. It was only a four-hour drive from Ithaca to our home in Northern New Jersey, and I drove home often enough—once a month or so—to keep up with things at home.

"I need to tell you something, and I need you to listen and stay on the line," my father said intently. "Your mother died today in a car accident."

I don't remember another word he said. I froze in place and fell to the floor. The next thing I remember is another call on that same pay phone. This time, I was right there to answer it. It was my brother Philip, a senior at Cornell University just five miles away in Ithaca.

"I'll come pick you up," he said matter-of-factly. "Be ready in an hour. We have to drive home."

It was late afternoon, and darkness was setting in. The snowfall persisted. I packed. Phil came up to the room to get me, and I tossed my bag into our old Volvo and tried to settle in for the long drive to Oak Ridge, New Jersey, through a whirling curtain of snow. I don't know how my brother drove through the darkness of that storm. We could barely see the road or the taillights of the cars in front of us.

We may have spoken, but I'm sure it wasn't much. I was lost in grief, and Phil was doing all he could to keep the car on the road. We made our way in a slow crawling caravan along the New York State Thruway. That typical four-hour trip took us eight treacherous hours. We arrived home at midnight.

Our neighborhood, always remote, was steeped in moonless darkness, and the swirling snow stubbornly persisted. My dad was awake, alone, and a blanket of heavy sadness enveloped us all. We hugged, but it was the arms down at the side kind of hug when you don't have the strength for anything else.

Looking back, that trip home was the first indication that, even in death, my mom would be watching over me. There's no other way to explain how we made it home safely that winter night in 1973. Or how I made it through the days that followed.

* * *

Throughout this book, I have written of success, failure, exhilarating highs, and excruciatingly desperate lows. I've written about all-consuming loves and about people whose manners and personalities I despised. Somewhere along the way, I learned how to deal with them—not always in stride but always with a dogged determination to survive, achieve, and recover.

I always keep two quotes mounted on my office wall to continually remind me to keep perspective. One is from Rudyard Kipling: "If you can meet with success and failure and treat them both as imposters, then you are a balanced man, my son." The other is from the thirteenth-century Persian poet Rumi: "Let yourself be silently drawn to the stronger pull of what you really love."

How some messages, quotes, and memories attach themselves and stay with us while thousands of others drift beyond our recollection is beyond me. It's not my place here to analyze or explain. I've often said that life experiences are a huge carousel spinning at light speed in our minds, stopping randomly to remind us of something either significant or not. As I wrote about my mother's death, that carousel stopped on a conversation I had with my high school sociology teacher, Mr. Richard O'Rourke.

At my mother's funeral, Mr. O'Rourke firmly grabbed my arm and pulled me to the side of the viewing room. He held me close with fiercely focused eyes just inches from mine.

"Mark, don't ask why," he said. "There is no answer to that question. So don't look or pray for one. Live on and live in her honor, with her grace."

He walked straight to the exit. I never saw him again.

* * *

I tell stories. I collect experiences, not money. Dollars sift through my hands like the sand slipping through my fingers on Long Beach Island when I was growing up. Foreshadowing? Ha!

Through all of this, a day doesn't go by without me thinking of my mother, reminding myself to find balance in my life and comfort in my self-belief. More than the next opportunity, I seek the next experience and a way to enrich the lives of others around me with whatever skills I have to offer. This is what I live for. This is my true love.

When you reach that level in your life, you learn to stop punishing yourself for what you perceive to be your mistakes and that neither your successes nor failures define you. In my case, I believe I can write a book and share all my experiences. My mom's influence is woven throughout this book, often without reference. My dad's stubbornness and resilience are present in this fabric as well.

Ultimately, what I have learned—and believe me, it took a while—is that there are very few absolutes. We waste so much time trying to make things fit or attempting to figure things out that aren't meant to be figured out. My parents were so different from each other, and their contributions to my life reflect these differences.

For too many years, I tried to make them both fit into a category that helped me deal with them in a way that made sense to me. But after so many pages and so many stories, I believe I have finally found my comfort with them. I've achieved so many of the things I've set out to achieve, and, while I fell short in some ways, I never cheated anyone or myself through my efforts. That's my dad.

I've always practiced compassion in my work and in my interactions with others, sometimes to a fault. From all my journeys, from all my experiences, the one constant is my belief in myself. It wavers at times, yet it stubbornly persists. That's my mom.

I simply would not have been able to do this without their unique contributions to my temperament and disposition. I am at peace with what I've accomplished and with what I have written. And like all things in life, I know there is more to come and more to achieve. What's next? Hmmmmm…

Acknowledgments

Yikes! Where do I start, and how do I determine this? No ground rules and no guidelines. So let's just get into it.

My Parents: Joseph Henry and Virginia Betty. Both are deceased and are so much a part of my fabric. Dad taught me the importance of hospitality, hard work, and how not to be a writer. Mom was just the most decent human being you'd ever want to meet. She was strong, resilient, smart, and patient but tough in her own way. She kept me humble and helped me believe in myself, no matter how awkward it got.

Vera M. Philipps: My high school English and writing teacher and mentor. She wrote in my yearbook: "The name Brezinski will always be a rare pleasure to me. I thank you for yourself." Mrs. Philipps taught me to write in my conversational style so that I wasn't always searching for words. She saw something in me that I didn't see in myself.

Coach Tom Brown: My high school basketball coach recognized my intense drive, a force that got me onto the varsity team—the only sophomore to attain such status. He believed in my competitiveness and commitment to succeed, which helped me develop my confidence to achieve. Under his tutelage, I made all-star teams and earned accolades for leadership I carried with me throughout my career.

Robert T. Gordon: My first real boss in the restaurant business and a genuine Texas character. He provided me with my first opportunity to coauthor a book—albeit a technical one—titled *The Complete Guide to Restaurant Management*. It helped keep my dream of writing alive. One Thanksgiving, he even set me up for a dinner date with Farrah Fawcett, whose parents were his and his wife Jean's neighbors in Champion Forest outside of Houston. Farrah ultimately canceled her Thanksgiving plans with me at the last minute to be on the set to film one of her movies. (This was around 1980 or so.) I ate alone that Thanksgiving. But it makes for a great story and moments to dream, "What if...?"

Parker Cowand: He believed in my vision for these stories and helped me get them over the goal line.

Mark Stuertz: A gem of a writer, my editor, friend, and one-time restaurant critic with a sharp tongue that never lies.

Shannan Metcalf-Perciballi: Another gem who never met an "i" she left undotted or a "t" she left uncrossed. She is a dear and loyal friend who made sure you didn't need a highlighter to note the mistakes.

Kenny Bowers: As good, generous, and genuine a friend as you'd ever want to have by your side in good times and in bad. As salt-of-the-earth as someone can be and a profoundly creative soul that never rests.

Asebe and Camila: My family for so many years, providing the moral support I needed just when I needed it, plus a push and a reason to carry on.

Anthony Tahmosh: He'd get mad at me if I gave him too many accolades, but without his support and belief in me, this book might never have been written. He is generous, selfless, and oh-so-smart.

My siblings: Dick, Lois, Doug (rest in peace, brother, and say hi to Mom and Dad), and Philip. Families are sensitive and fragile things, and ours was no different. But through it all, we all still talk;

maybe not as much as we should, but we're never too distant from each other when in need.

Fellow restaurateurs and friends: There are too many to name. But we are truly a quilt of varied colors and fabrics, and the essence of what we do is to offer tasty choices for our neighbors and friends. It's that choice we strive for and thrive on, and it is what drives us. I've been blessed to be there to see so much of our progress and have enjoyed all of you and all of this. I hope that, in some way, I have shared my appreciation of you, but if I haven't, hit me up for a meal or drink next time you see me. I'll take the check, please.

And finally: To every guest who ever walked into any of the restaurants that I had a hand in creating or working in. You are the lifeblood of our business, and all the good people I know and/or have mentioned in this book are dependent on your support and satisfaction with what we create daily. It is inarguable that sometimes we lose our way and don't put our best foot forward.

But I promise you it's never because we don't give a shit. We have our bad days too. But no one—or almost no one—I know gets into the restaurant business to simply fleece people out of their hard-earned money. You folks are too smart to put up with that kind of nonsense, and those sorts of places don't last long. You may not like or understand some restaurants, but I can assure you most operators and places are striving to win you over and praying you'll become regulars.

Please keep giving them a chance to do just that. And please support the ones that do it better and give you the nourishment you seek—not just for your belly but for the part of you that needs a break from the day-to-day in a comfortable, friendly, and visually appealing environment. It's what we do and is at the heart of why we choose this business.

About the Author

Mark H. Brezinski is a forty-five-year veteran of the dynamic restaurant industry. A graduate of the famed Cornell University of Hotel Administration, Mark went on to create and co-create multiple nationally acclaimed restaurant companies and concepts. His lively experiences in search of unique culinary discoveries span the globe. His connections to industry icons like the late Norman Brinker, Paul Fleming, Phil Romano, and Chefs Mark Miller and Michael Mina inspire stories that enlighten and entertain.

His restaurants have featured an international buffet of flavors that include Indian, Italian, French, and Vietnamese. His travels have taken him from Mumbai to Maui, Tokyo to Tuscany, and San Francisco to Shanghai. This collection of tales takes you to all these storied destinations on journeys that are sometimes funny, often heart-breaking, and never without lessons. This log of expeditions is a remarkable tag-along to the highs and lows of entrepreneurship and the self-discovery that comes with each stop.

Mark resides in Dallas and is never one to leave a stone unturned. He is currently launching his latest concept, Bizzy, his interpretation of better fast food.

About the Co-Author

A James Beard award-winning journalist, Mark Stuertz has been a Dallas-based writer and author for more than twenty-five years. His writing has appeared in a variety of publications, including the *Dallas Observer, Texas Monthly, American Way, Spirit, Food & Wine,* and *American Driver.* He is the author of *Secret Dallas: A Guide to the Weird, Wonderful, and Obscure,* editor of the *Fearless Critic Dallas Restaurant Guide,* and co-author of *Dream Makers* with Jim "The Rookie" Morris, and *From Homeless to $100 Million* with Ron Sturgeon.